THE OPEN UNIVERSITY

Science: A Second Level Course

Structure, Bonding and the Periodic Law

8 Trends in the Periodic Table; Orbitals Part 1

9 Elements of the Lithium Row; Orbitals Part 2

LIHE STACK

Prepared by an Open University Course Team

The Open University Press

A NOTE ABOUT AUTHORSHIP OF THIS TEXT

This text is one of a series that forms part of a Second Level Science Course, *Structure, Bonding and the Periodic Law*. The other components are a series of television and radio programmes, home experiments and a summer school.

The Course has been produced by a team, which accepts responsibility for the Course as a whole and for each of its components.

THE S25- COURSE TEAM

Chairman and General Editor
L. J. Haynes

Unit Authors
C. J. Harding
D. A. Johnson
Joan Mason
Jane Nelson
R. A. Ross

Editors
Jacqueline Stewart *(Editor)*
Maggie Harris *(Editorial Assistant)*

Other Members
A. Clow *(BBC)*
R. M. Haines *(Staff Tutor)*
R. R. Hill
D. S. Jackson *(BBC)*
G. W. Loveday *(Staff Tutor)*
G. D. Moss *(IET)*
D. R. Roberts
R. C. Russell *(Staff Tutor)*
N. A. Taylor *(BBC)*
Christina Warr *(Course Assistant)*
B. G. Whatley *(BBC)*

The Open University Press
Walton Hall, Bletchley, Bucks.

Copyright © 1973 The Open University.
First published 1972 ; 2nd edition 1973.

Designed by the Media Development Group of the Open University.

Printed in Great Britain by
Martin Cadbury Printing Group, Cheltenham and London.

ISBN 0 335 02262 6

This text forms part of an Open University Second Level Course. The complete list of Units in the Course is given at the end of this text.

For general availability of this text and supporting material, please write to the Director of Marketing, The Open University, Walton Hall, Bletchley, Bucks.

Further information on Open University courses may be obtained from the Admissions Office, The Open University, P.O. Box 48, Bletchley, Bucks.

2.1

Unit 8 Trends in the Periodic Table; Orbitals Part 1

Contents

Objectives

When you have finished this Unit you should be able to:

1   Define or recognize definitions of, and recognize correct uses of the terms and principles listed in Table A (SAQs 1, 8, 9, 14).

2   Recognize representations of, or represent s and p atomic orbitals using orbital boundary surfaces (SAQ 3).

3   Relate s and p orbital occupancy to the energy and distribution of electrons in atoms (SAQ 1).

4   Given an appropriate energy level diagram, apply the aufbau procedure (including Hund's rule and Pauli's principle) to determine s and p orbital occupancy in atoms and simple molecules, and decide whether the molecules are stable (relative to gaseous atoms), and what bond orders are predicted (SAQs 2, 3, 6).

5   Use the criterion of unpaired electrons to decide whether an atom or molecule is paramagnetic (SAQ 4).

6   Apply simple molecular orbital predictions to determine the shapes of simple hydrides and compare these predictions with those of electron pair repulsion theory (SAQs 7, 9).

7   Apply simple symmetry arguments to decide whether linear combinations of given s and p atomic orbitals produce σ or π molecular orbitals, or no orbital (SAQ 5).

8   Show how the electronegativity of an element is described and estimated (SAQs 10–12).

9   Define and use the following measures of atomic size:
(a)   the covalent radius;
(b)   the ionic radius (described in Unit 7);
(c)   the non-bonded (van der Waals) radius (SAQs 12, 13).

10   Describe the periodicity of the following properties, and explain this qualitatively in terms of energy shells of electrons in atoms, and the effective nuclear charge:
(a)   ionization energy;
(b)   electron affinity;
(c)   electronegativity;
(d)   the covalent radius;
(e)   the ionic radius (SAQs 10–13, 19–21).

11   Describe and account for the distinguishing characteristics of the different kinds of hydride (SAQs 14–17).

12   Describe the origins and consequences of hydrogen bonding in water (SAQs 15, 16).

4

Table A

List of scientific terms, concepts and principles used in Unit 8

Introduced in a previous Unit	Unit Section No.	Developed in this Unit	Page No.
	S100*	antibonding orbitals	16
atomic spectra	6.4	atomic orbital	8
Coulomb law	4.2.2	aufbau procedure	11
degeneracy	30.3.1	bonding orbital	16
electron energy shells	7.1.2	boundary surface	9
electron energy subshells	7.2	core	32
electron spin	7.3	covalent radius	34
electronic configuration	7.2.2	effective atomic number	32
Heisenberg's uncertainty principle	29.4	effective nuclear charge	32
Hund's rule	7.4	electron correlation	14
hydrogen bonding	10.5.2	free radical	36
kinetic energy	30.4.2	inductive effect	33
nodal surface	30.4.3	ligand	34
pairing of electron spins	7.3	linear combination of atomic orbitals	15
Pauli exclusion principle	7, App 2	molecular orbital	15
potential energy	4.4.3	non-bonding orbital	25
probability wave	29.3	orbital hybridization	23
quantum numbers n, l, m, s	7, App 2	orbital overlap	15
radial probability wave function	30.5.2	paramagnetism	52
s, p, d, nomenclature	7.2.2	post-transition elements	39
unpaired electrons	7, App 2	self-consistent field calculation	12
	S25-	sp³ hybrid orbital	23
electron affinity	5.8	symmetry of orbitals	16
electronegativity	5, App 1		
ionic radius	7.1.5		
bond dissociation energy	**Data Book****, Section 8		

* The Open University (1971) S100 *Science: A Foundation Course*, The Open University Press.

** The Open University (1973) S24-/S25- *The Open University Chemistry Data Book*, The Open University Press.

Study guide

In this Unit we introduce molecular orbital theory, an extension of wave theory, to the behaviour of electrons in molecules. Consequently, we start by examining how wave theory describes electrons in atoms. Atomic structure and electron waves in atoms are subjects that you have already met in S100, Units 7 and 30; so, perhaps more than in any other Unit of this Course, we depend on your knowledge of S100. In particular, we assume that you are familiar with the ideas developed in Unit 7 and in Unit 30, Section 30.5.

Because an understanding of these ideas is crucial to what follows, we begin this Unit with an exercise based on the concepts in S100. Even if you are able to answer them, you should read the answers and comments to these questions. You should also read S100, Unit 7, Appendix 2 (Black) about the position of the electron, before starting to study this Unit. You will find that this Appendix helps to relate the experimental observations discussed in S100, Unit 7, to the theoretical approach of S100, Unit 30.

Although a quantitative application of molecular orbital theory involves mathematics outside the scope of this Course, it is possible to apply the concepts of the theory in a completely qualitative way. You will need almost no mathematics to study this Unit.

8.0 Introduction

In Unit 7 of this Course we examined critically the electron repulsion theory of chemical bonding. Evidently there are many molecules not adequately explained by this theory.

You have already encountered the basis of another bonding theory. In S100, Unit 30, the behaviour of electrons in atoms was described by a wave equation, the Schrödinger equation.

In the first part of this Unit we examine the application of wave theory, or wave mechanics, to atoms and simple molecules. Often simple qualitative arguments allow us to predict the shapes and stabilities of molecules, and we compare these with predictions of the more elementary theory that you met in Unit 7.

We proceed then to a study of the Periodic Table as a prelude to the later part of this Course, which is concerned mainly with the non-metals. If we ask ourselves how and why non-metals differ from the metals, the answer is bound up with the Periodic law. By studying the progressions (from metal to non-metal) in the Periods and Groups, we find that simple relationships of atomic structure can help us to understand the generalizations, and many of the singularities, of inorganic chemistry.

We conclude this Unit with some descriptive chemistry of hydrogen, which affords useful illustrations both of our bonding theory, and of the trends in the Periodic Table that we describe earlier in the Unit.

8.1 The hydrogen atom

Exercise

1 What information about the electron in the hydrogen atom can be obtained by studying the flame (atomic) spectrum of hydrogen?

2 What evidence is there to support the notion that electrons in other atoms exist in specific energy levels?

3 What experimental evidence is there to show that a beam of electrons has wave properties?

4 You will recall from S100, Unit 30, that a string which is clamped at both ends can vibrate only with certain frequencies which are determined by a quantum number. A two-dimensional membrane is also restricted in its vibrations which are defined by two quantum numbers. How many quantum numbers would you expect to apply to an electron constrained to move within the electric field of a nucleus?

The answer to this exercise is on p. 52. Remember, even if you can answer all the questions, you should read the answers and comments before going on.

The first successful application of wave mechanics to an atom was made in 1926 when Schrödinger solved the wave equation for the hydrogen atom. The solution provides two properties of the electron in the atom, its energy and its distribution.

Wave theory depicts the electron as a three-dimensional wave, and it is reasonable to expect the solution of the wave equation to depend on three quantum numbers (S100, Unit 30, Section 5). These are identical to the quantum numbers used to specify energy levels in S100, Unit 7:

the principal quantum number n
the azimuthal quantum number l
the magnetic quantum number m

The values that n, l and m can take are restricted by the rules given in S100, Unit 7.

n can take only integral values; 1, 2, 3 . . .
l can be 0, 1, 2 . . . up to $n - 1$
m can take integral values from $+l$ to $-l$

7

Energy

An energy of the electron can be calculated for each set of values of n, l and m. The calculated energy levels are shown schematically in Figure 1, and these energies are in exact agreement with the energies given by the spectrum of atomic hydrogen. Notice that the energy depends only on the value of n. Except for the $n = 1$ level, the energy levels in the hydrogen atom are each degenerate (the energy levels are each associated with more than one state).

Remember that the energy depicted in Figure 1 increases upwards towards more positive values. Zero energy is above the top of the Figure, so that all the energy levels have negative energies. The more negative energy levels correspond to greater stability because more energy is needed to remove an electron in the $n = 1$ level than the $n = 2$ level, i.e. more energy is needed to excite an electron from the $n = 1$ level to zero than from the $n = 2$ level to zero. Throughout this Unit this is the convention we use in the energy level diagrams.

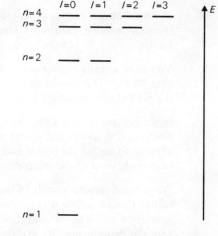

Figure 1   Energy levels in hydrogen.

Distribution

The second property we can calculate is the wave function ψ and the way ψ varies in space. From this we can determine the electron density or probability, ψ^2, of finding the electron at any particular point in space. We call the solutions (for ψ) of the Schrödinger equation *orbitals*—and we say that the electron occupies an orbital. This does not mean that the electron moves in a particular orbit, but the term is a legacy of an earlier theory in which the electron *was* pictured as moving in a defined orbit around the nucleus, a description which violates Heisenberg's uncertainty principle (S100, Unit 29, Section 29.4). So ψ is the amplitude of the wave, but like other waves, the intensity is ψ^2. ψ^2 can be interpreted in two ways: if we consider the electron as a wave, then ψ^2 is the electron density in a given region; if we consider the electron as a particle, which nevertheless obeys Heisenberg's uncertainty principle, then ψ^2 is the probability of finding the electron in any particular region of space. The values of ψ^2 in the region around the nucleus where the electron is most likely to be found give us an impression of the electron distribution in the orbital.

For the hydrogen atom with its single electron this impression tells us the shape and size of the atom. The nature of the Schrödinger equation for the hydrogen atom is such that ψ can be factorized into two parts.

$$\psi = \psi_r \cdot \psi_a$$

This factorization is conceptually very convenient. ψ_r is the *radial function* and contains only the information of how ψ varies with the distance r from the nucleus. ψ_a is called the *angular function* and tells us the shape of the orbital. We shall discuss this shortly. ψ_r is related to a property called the radial probability, P_r (S100, Unit 30, Section 30.5.2).

$$P_r = 4\pi r^2 \, \psi_r^2$$

The radial probability answers the question: at what distance from the nucleus in any direction are we most likely to find the electron? So ψ_r tells us the size of the orbital. Figure 2 shows how ψ_r and P_r depend on r for various states of the hydrogen atom. The functions are labelled according to the notation used for the electronic configuration of atoms (S100, Unit 7).

Figure 2 shows that in the 1s state the electron is likely to be in a region very close to the nucleus. In the 2s state a node ($\psi = 0$) occurs in the radial function, so there are two regions where the electron is most likely to be found or where the electron density is high.

> What relation do you see between the radial probabilities and the quantum numbers n and l?

As n increases, so the maximum in the P_r curves moves further from the nucleus. Although the probability reaches zero only at an infinite distance from the nucleus, the P_r curves give us some idea of atomic size. The hydrogen atom is bigger in an excited state ($n > 1$) than in the ground state ($n = 1$). You might also notice that the number of nodes (positions where $\psi = 0$) increases both with n and l; for s orbitals there are $n - 1$ nodes, for p orbitals $n - 2$.

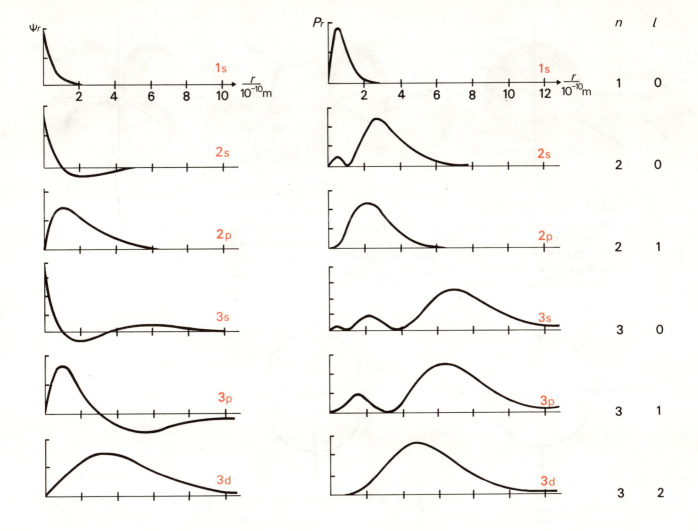

Figure 2 *Radial functions and radial probabilities.*

The angular functions ψ_a give the shapes of the orbitals by telling us how ψ varies with the direction from the nucleus.

For an s orbital

$$\psi_a = 1$$

Thus the wave function ψ is determined only by the value of the radial function ψ_r. So at a distance r from the nucleus, ψ has the same value in all directions. In other words, s orbitals are spherical.

The angular function for one of the 2p orbitals is given by

$$\psi_a = \frac{\sqrt{3}}{2\sqrt{\pi}} \cos \theta$$

where θ is the angle between the vertical axis and the direction in which ψ_a is calculated. Obviously ψ_a now depends on the direction from the nucleus. The result of this is that 2p orbitals have a nonspherical shape. Rather than attempt to represent the angular functions here, we shall proceed with a representation of the total wave function ψ. You may well encounter representations of angular functions in your reading, so we have included one of these representations in Appendix 3 (Black).

We can represent orbitals with *boundary surfaces*, as in Figure 3. These are surfaces of constant ψ value, where ψ has a higher value inside the surface than outside. In other words, they are contour diagrams and give us a picture of the total wave function. The s orbitals are spherical. The 2p orbitals have two lobes where ψ has a positive value* in one lobe and a negative value* in the other.

---

* The signs of the value of ψ are indicated in Figure 3 and subsequent figures. They show where ψ has a positive amplitude and a negative amplitude. These signs do not represent electric charges.

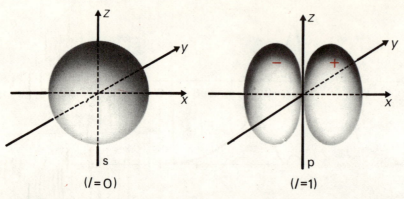
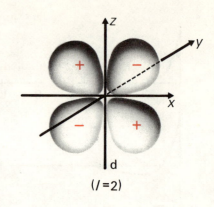

(/ = 0)   (/ = 1)   (/ = 2)

Figure 3   Orbital boundary surfaces.

There are three p orbitals, each directed along one of the perpendicular axes x, y and z. The d orbitals are more complex, but we shall not use them in our present treatment. For our purposes, it is not necessary to represent these s and p orbitals in three dimensions and we shall use schematic diagrams as shown in Figure 4. These diagrams are merely two-dimensional outlines of the orbital boundary surfaces, and Figure 4 shows an s orbital and a $2p_z$ orbital.

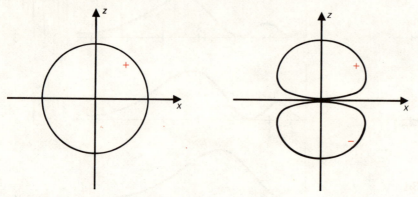

Figure 4   Schematic representation of orbitals.

Do the orbital boundary surfaces in Figure 3 also represent the electron distribution?

If they are surfaces of constant ψ, then ψ^2 must also have a constant value on the surface. They are therefore surfaces that enclose a certain fraction of the electron distribution or charge cloud.

So an atomic orbital is represented by a boundary surface joining points with the same value of ψ. This value is chosen so that some fraction (often 0.95) of the total electron distribution ψ^2 is inside the surface. Later, in Section 8.4, we shall see how important the shapes of the atomic orbitals are in the geometry of molecules.

SAQ 1 (Objectives 1, 3)   How do (a) the energy and (b) the distribution of an electron in an orbital of the hydrogen atom depend on the two quantum numbers n and l?

Now summarize what you consider to be the main points of this Section and compare your summary with that given below.

Summary

1   The wave description of electrons in atoms leads to a model of the atom in which the electron occupies an orbital (a region of space defined by the solution of the wave equation).

2   This treatment gives energy levels in the hydrogen atom in exact agreement with experiment.

3   The wave equation can be factorized into two parts: a radial function which tells us the size of the orbital, and an angular function which tells us the shape of the orbital.

4   We can represent orbitals by orbital boundary surfaces or schematic outline diagrams.

8.2 More complex atoms

In the previous Section we considered a very simple system: a nucleus and one electron. A single interaction is involved and Schrödinger was able to solve the wave equation exactly. Once the number of electrons increases, the problem becomes much more complex. The motion and energy of each electron depends on both the nucleus and all the other electrons.

So each electron interacts with the nucleus and with each of the other electrons. Helium, for example, has two electrons, and so there are three interactions. When we get to carbon, with 6 electrons, there are 21 interactions to consider. In fact, the wave equation cannot be solved exactly for atoms more complex than hydrogen. However, chemists in the period around 1930 were spurred on by the success achieved for the hydrogen atom (the precise calculation of energy levels) and accepted that even inexact or approximate solutions might prove useful. What sort of approximations should we make? Here we are helped by experimental evidence.

Experimental evidence

The atomic spectra of the alkali metals are broadly similar to the spectrum of hydrogen. Of course, the spectra are more complex, but the similarity suggests that the sequence of energy levels in the alkali metals is roughly similar to that in hydrogen. The spectra of other elements turn out to be much more complex, but there is other evidence to suggest that the energy levels in other atoms are also similar to those in hydrogen.

The sequence of successive ionization energies for sodium, which was discussed in S100, Unit 7, were explained by the existence of 1s, 2s, 2p and 3s electron sub-shells. Other elements show a similar sequence of ionization energies.

All this suggests that the wave equations for electrons in heavier atoms are roughly similar to those for hydrogen. We, therefore, begin with a set of wave functions like those for hydrogen and try to improve them.

Before we can do this we need to know how the electrons occupy the orbitals, and here we turn to the empirical rules introduced in S100, Unit 7.

The aufbau procedure

In S100, Unit 7, electrons were said to occupy shells or sub-shells and you applied a set of rules to determine which shells were occupied. In wave theory, the electrons occupy orbitals and the translation of the rules into this new terminology is as follows.

1   We feed the electrons into the orbitals. Each electron goes into the lowest energy orbital to give the lowest energy or ground state of the atom.

2   The number of electrons that can occupy one orbital is limited by the *Pauli exclusion principle*. This says that no two electrons can have the same four quantum numbers. To put it another way, a maximum of two electrons can occupy any one orbital and then they must have their spins opposed thus ↑ and ↓. Pauli's principle is essentially an empirical formulation based on the study of atomic spectra and the atomic structure and periodicity outlined in S100, Unit 7. It amounts to saying that two electrons can occupy the same region of space (have the same ψ function) only when their spins are opposed.

Pauli principle

3   When we encounter degenerate orbitals, i.e. orbitals in which the electrons have the same energy, we apply *Hund's rule* which says that there is always a maximum number of unpaired electrons consistent with rule 1. This result is hardly surprising. Effectively it says that, as long as there are no other energetic factors to consider, electrons tend to avoid being in the same region of space.

Hund's rule

Using these rules, we determine which orbitals are occupied, i.e. we determine the electronic state of the atom. This is called the *aufbau* (building up) *process*. It is important to realize that this is not a physical process, but part of the procedure we employ to build up a model of an atom. Once we have established a sequence of wave equations (orbitals) of increasing energy, we assign electrons to the orbitals.

Now let us turn to the problem of solving the Schrödinger equation for a many-electron atom. We know that the equation can be solved for a single electron moving in the spherical electrical field of the nucleus*. The following procedure which reduces the calculation to a number of steps each like that for hydrogen was suggested by Hartree. So we single out one electron to begin our calculation and assume that we can average the effect of all the other electrons.

We assume that the other electrons have hydrogen-like wave functions**. This allows us to calculate their distribution and hence the average potential field affecting our single electron. If the electrons occupy only s orbitals, this potential field is spherical. But if the other electrons occupy p or d orbitals the field is non-spherical. We then have to average the field over all angles to give an average spherical field. With this information, the Schrödinger equation for our single electron can be solved. We now choose a second electron and repeat the process, this time including the new wave function for the first electron in the calculation of the average electric field. And so on for each electron. We end up with a new set of wave functions which we can use as a starting point for a complete new cycle of calculations, and so on. Provided that the first approximate wave functions were not wildly in error, this type of iterative procedure is convergent; it reaches the stage where there is no significant change in the potential fields in one cycle. We call the operation a *self-consistent field calculation*, and it is just this kind of laborious repetitive process that can be tackled with a computer.

self-consistent field calculation

This is just one of the methods used to calculate atomic orbitals. A more empirical one (Slater's method) is mentioned in Appendix 5 (Black) (p. 50).

We began with a set of hydrogen-like wave functions and treated the atom as a single-electron problem at each stage. The importance of this is that the improved angular functions turn out to be similar to those for hydrogen. This being so, we can continue to represent the orbitals with the diagrams in Section 8.1 (see Figs. 3 and 4).

Now, let us examine some of the results of these calculations.

Comparison of theory and experiment

There is no experimental evidence that the distribution of electrons in atoms varies with direction, so we cannot examine the *angular* distribution of electrons.

We *can* make a comparison of the radial distribution of electrons. For gaseous atoms, this can be measured experimentally by electron diffraction. In Unit 2 you learnt that a beam of electrons is scattered by the nuclei in atoms and molecules. The electron beam is also scattered by the electrons in the atom. Figure 5 is a comparison of the results for argon, showing the theoretical and experimental distribution of electrons. A striking feature is the clear separation of the electrons into 'shells' corresponding to values of the quantum number n.

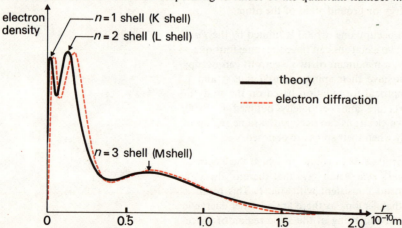

Figure 5   *Radial electron distribution in argon.*

* A spherical field is one in which the strength of the field at a particular distance from the nucleus is constant irrespective of the direction. An electric charge at a point produces such a field.

** Wave functions that describe one electron moving in an electric field similar to that in a hydrogen atom.

The important difference between the orbitals of hydrogen and those of many-electron atoms lies in their energies. For example, in many-electron atoms the radial distribution of electrons for the 2s and 2p orbitals is different; the electron in the 2s orbital penetrates closer to the nucleus and so its energy is lower. The 2s and 2p orbitals are no longer degenerate; they no longer have the same energy. This result is quite general. Except in hydrogen, orbitals with the same value of n but different values of l have different energies. The sequence of energy levels which emerges from these calculations is:

$$1s < 2s < 2p < 3s < 3p < 4s \approx 3d$$

and this is confirmed by atomic spectroscopy.

That is, electrons in atoms occupy orbitals which have characteristic energies. More energy is required to remove an electron from the lowest energy orbital (1s) than from higher energy orbitals. For example, in sodium the easiest electron to remove is in the 3s orbital (the valence shell electron).

We can represent these ionization energies by a diagram which shows the energy required to remove an electron from each orbital in the atom (Fig. 6). This energy scale in Figure 6 is purely arbitrary and non-linear; the figure merely shows the sequence of energies. Each box represents an orbital which can hold up to two electrons.

Figure 6   Energy level diagram.

When one electron is removed from the atom, the energies of all the other electrons are affected, and the diagram will be different for the removal of a second electron. Consequently, the total energy of the electrons in the atom cannot be determined by summing the energies in Figure 6.

Now, compare the sequence of energy levels shown in Figure 6 with that for hydrogen shown in Figure 1.

The contrasting feature is the separation of the sub-shells, the s, p and d levels. In hydrogen these levels are degenerate but in atoms with more than one electron the levels have different energies. The difference between 3p and 3d is so great that the 4s level is actually lower than the 3d level in Figure 6. We stressed that the energy scale in Figure 6 is purely arbitrary; obviously the energies will change from element to element. In fact the separation of the sub-shells increases with the atomic mass and the 3d level is lower than 4s for very light elements up to nitrogen in the Periodic Table.

Chemical periodicity also supports the results of the calculated energy levels.

Figure 6 shows quite clearly the emergence of the pattern of the Periodic Table:
  the first period of two elements, H and He;
  the second period of eight elements, Li to Ne;
  the third period of eight elements, Na to Ar;
  the fourth period of eighteen elements, K to Kr.

Just as with the hydrogen atom, we should hope to be able to calculate the energy levels accurately, i.e. so that they are in agreement with spectroscopic data. Simple self-consistent field calculations give the energy of the helium atom to within about 1 per cent. However, the method involves one severe assumption: when the wave function of the second electron is calculated, the effect of the first is averaged over the whole atom.

For a moment, let us suppose that the electrons behave like particles and that we can be fairly certain of their instantaneous position.

If one electron is on one side of the nucleus, where would you expect the second electron to be?

Because the two electrons repel each other, the second electron will probably be on the other side. The position of each electron depends on the *instantaneous* and not the *average* position of the other electron. So, by assuming an average position for each electron, we overestimate the repulsion between the electrons. The difference between the energy calculated by the self-consistent field method and that of an exact solution is called the *correlation energy*. To allow for this we introduce into the calculation a dependence on the interelectronic distances. When this is done for the helium atom, the experimental and calculated energies differ insignificantly (about 1 in 10^5). Agreement of this kind is sufficiently precise to be called exact and it has encouraged chemists to apply the theory to more complex systems. However, allowing for electron correlation becomes increasingly more difficult with larger molecules and is only attempted in very sophisticated calculations.

electron correlation

Now summarize what you consider to be the main points in this Section and compare your summary with the one below.

Summary

1   The spectrum of the alkali metals and the sequence of successive ionization energies of atoms suggest that the energy levels in many-electron atoms are similar in pattern to those in hydrogen. This in turn suggests that the atomic orbitals (wave functions) are also similar.

2   Beginning with a set of atomic orbitals like those for hydrogen, we assign electrons to the orbitals according to certain rules. We build up a picture of the atom with the aufbau procedure:

(a)   electrons occupy the lowest energy orbitals available;

(b)   Pauli's principle allows us to put not more than two electrons (with opposite spins) in one orbital;

(c)   when degenerate orbitals are available, electrons occupy different degenerate orbitals singly in preference to pairing in one orbital (Hund's rule).

3   The set of wave functions is then improved one at a time by the self-consistent field method. Sophisticated calculations allow for the correlation of electrons.

4   Radial distributions and energies calculated for electrons in light atoms are in excellent agreement with the results of experiments. The sequence of energy levels also agrees with the periodic sequence of elements.

SAQ 2 (Objective 4)   The oxygen atom has eight electrons. Use the energy level diagram (Fig. 6) to determine which orbitals are occupied in the ground state of oxygen. What criteria do you need to use to decide on the allocation of electrons?

SAQ 3 (Objectives 2 and 4)   On the basis of orbital occupancy, which of the atoms He, Li and O are predicted to have spherical electron distributions?

SAQ 4 (Objective 5)   Which of the atoms in SAQ 3 would you expect to be paramagnetic?

8.3   Molecular orbitals, H_2 and He_2

Now, let us see if the ideas developed so far can be applied to molecules. We begin by considering a very simple molecule, H_2.

Under normal conditions, hydrogen gas exists as diatomic molecules, H_2, i.e. the hydrogen molecule is stable with respect to hydrogen atoms. This stability can be expressed quantitatively by, for example, the Morse curve (S100, Unit 11), which shows how the energy of the molecule depends on the internuclear separation (Fig. 7).

We know the details of this curve from the spectrum of H_2. The important features are:

the molar bond dissociation energy D_m (H — H) = 436 kJ mol^{-1}
the equilibrium bond length r_e = 74.1 pm.

Before we attempt to apply wave theory to the hydrogen molecule, let us stop to think about the distances and forces involved. Suppose two hydrogen atoms are placed at a distance of 74.1 pm apart. What can you say about the electrostatic forces involved between nuclei and electrons? (Look at Figure 2 to get some idea of the electron distribution in a ground state (1s) hydrogen atom.)

At this separation the two orbitals (i.e. the two electron clouds) overlap. The electron of one atom is therefore attracted by the nucleus of the other. Of course, there are also repulsive forces to consider, but here we can see the origin of a binding force in the molecule.

What then might we ask of a theoretical description of the hydrogen molecule? Certainly the molecule should be stable with respect to the atoms. The success achieved in calculations for atoms might tempt us to expect quantitative agreement with the dissociation energy and bond length.

Before we become too optimistic let us consider the magnitude of the binding energy in the hydrogen molecule. This, after all, is the quantity that tells us the stability of the molecule with respect to its constituent atoms.

The hydrogen molecule is more stable than two hydrogen atoms in spite of the repulsive forces between the nuclei. The energy of the electron in a hydrogen atom is 1314 kJ mol^{-1} (the ionization energy). So the energy of the electrons in the hydrogen molecule must be greater than twice this value. The binding energy of the molecule is only a fraction of the total electronic energy of the molecule. Therefore the task of calculating the bond energy accurately is formidable.

Now let us return to the problem of applying wave theory to the electrons in the hydrogen molecule.

Conceivably we might expect the procedure to be similar to that for atoms. The electrons occupy orbitals (now called *molecular orbitals*) which are described by wave functions. These are obtained from an approximate solution of the Schrödinger equation. Following the rules outlined in the aufbau process, the orbitals are occupied by the available electrons.

molecular orbitals

However, there is an important difference between an atomic and a molecular orbital. The molecular orbital describes the electron distribution in the molecule; it is a multi-centre orbital. That is to say, the orbital is centred on more than one nucleus. Obviously this makes the calculation more difficult and again we resort to approximation, this time somewhat more drastic.

Some idea of the approximation that is made can be gained by an intuitive consideration of the electrons' behaviour.

What forces will predominate when the electron is near one of the nuclei?

In this region, the electron is influenced almost solely by the electric field of that nucleus, and so it will be described quite well by the *atomic* wave function. When it is between the nuclei, an electron is influenced by both nuclei and so might be described by some combination of the atomic orbitals. This, together with the test of comparison of experimental and calculated results, is a justification of the procedure.

The two atomic wave functions for the atoms A and B are added together to give a molecular wave function. We can illustrate this graphically in Figure 8. We call this addition the *linear combination of atomic orbitals* (LCAO).

linear combination of atomic orbitals

The molecular wave function is given by:

$$\psi = N(\psi_A + \psi_B)$$

where N (*the normalization constant*) is introduced to ensure that the total electron probability summed over all space is exactly unity for one electron.

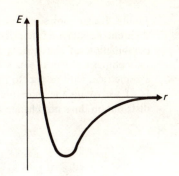

Figure 7   The Morse curve.

15

Figure 8 also shows a boundary surface of the molecular orbital. We use a different notation to describe the electronic configuration of molecules. This combination of the two 1s atomic orbitals gives the σ1s molecular orbital (sometimes written as sσ). σ denotes the *symmetry of the orbital*; a σ orbital is symmetrical with respect to rotation about the line joining the two nuclei. That is, if we rotate the molecule about this line, the appearance of the electron distribution does not change.

σ orbital symmetry

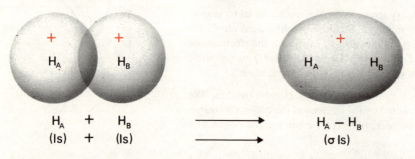

$$H_A + H_B \longrightarrow H_A - H_B$$
$$(1s) + (1s) \longrightarrow (\sigma\,1s)$$

Figure 8 The linear combination of atomic orbitals.

Notice that the two 1s atomic orbitals also have this same property of symmetry: they are symmetrical with respect to rotation about the line joining the two nuclei. We shall see shortly that symmetry plays an important role in molecular orbital theory.

In the formation of the σ1s orbital the linear combination gives an increase of electron density in the region between the two nuclei where the electron attracts both nuclei, and where it is attracted by both nuclei. So simple electrostatic forces can account for a decrease in the energy of the electron when the molecule is formed. This type of orbital, which draws the nuclei together, is called a *bonding orbital*.

Two atomic orbitals were combined to produce one molecular orbital. Each atomic orbital can hold two electrons (although in hydrogen only one electron is present); so we might ask whether the molecular orbital can hold four electrons, or is there some other way of combining the two atomic orbitals?

The Pauli principle is quite general, it applies to molecular orbitals. Two molecular orbitals can be generated from two atomic orbitals, the second by combining the two wave functions of opposite sign:

$$\psi = N(\psi_A - \psi_B)$$

This operation is shown graphically in Figure 9. Notice that in this molecular orbital there is nodal plane ($\psi = 0$) midway between the nuclei, where ψ changes sign. The electron density on the opposite side of the atoms to the bond is now increased.

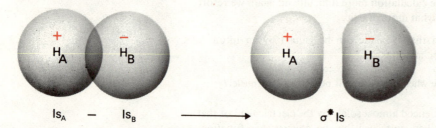

$$1s_A - 1s_B \longrightarrow \sigma^* 1s$$

Figure 9 The linear combination of atomic orbitals

What contribution would you expect an electron in this orbital to make to the stability of the molecule?

Formation of the molecular orbital produces a *decrease* in electron density between the nuclei. Consequently, an electron in this orbital tends to pull the two nuclei apart and make the molecule *less* stable. We call an orbital with this property an *antibonding orbital* and denote it by an asterisk.

Is this orbital a σ orbital?

If it is formed from the spherically symmetrical 1s orbitals, then it must have σ symmetry. So we denote the orbital by σ^*1s. The 1s again refers to the parent atomic orbitals.

16

While the $\sigma 1s$ orbital involves a lowering of energy (a more stable molecule), σ^*1s involves an energy increase (a less stable molecule). Figure 10 shows the formation of two molecular orbitals from the parent atomic orbitals, in terms of the energy changes.

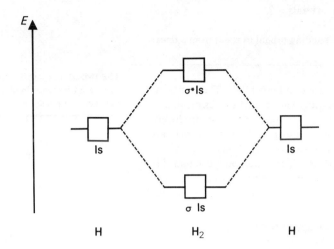

Figure 10   Energy level diagram for H_2.

In Figure 10 the energies of the electrons in the two atomic orbitals are represented on the left- and right-hand sides. The molecular orbital energies are shown in the centre of the Figure, $\sigma 1s$ at the bottom and σ^*1s at the top.

You should now be able to carry out the aufbau process, using the rules outlined in the previous Section, together with the energy level diagram (Fig. 10).

What is the orbital occupancy of the hydrogen molecule?

The hydrogen molecule has two electrons and, according to the Pauli exclusion principle, these both occupy the $\sigma 1s$ orbital with their spins opposed, thus $\boxed{\uparrow\;\downarrow}$. In the shorthand notation used for atomic orbitals (shells and sub-shells) in S100, Unit 7, the electronic configuration of the hydrogen molecule is $(\sigma 1s)^2$ where the superscript denotes the number of electrons in the orbital.

The hydrogen molecule, therefore, has two electrons in a bonding orbital. This means that there is a single bond between the two atoms, i.e. the bond order is one. (The bond order is equal to half the number of bonding electrons.)

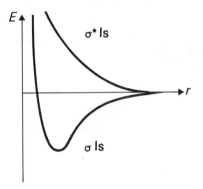

Figure 11   Potential energy curves.

Calculations involving the simple linear combination of wave functions are only approximate. For many-electron atoms, accurate results are only obtained when *electron correlation* is introduced. Similarly, we do not expect the two electrons in the hydrogen molecule to occupy the same region of space simultaneously. When electron correlation is introduced, agreement between theory and experiment is within the limits of experimental error for the hydrogen molecule.

The Morse curve (see Fig. 7, p. 15), gives the variation of energy with internuclear distance when the hydrogen molecule is in its ground state $(\sigma 1s)^2$. When the electrons occupy the antibonding orbital the molecule is unstable with respect to dissociation into atoms (Fig. 11).

The helium atom contains two electrons, both in the $\sigma 1s$ orbital. Linear combination of these orbitals gives an energy level diagram that is similar to Figure 10, although the energy levels have different numerical values.

Represent the orbital occupancy of He_2 in Figure 12 with arrows (the aufbau procedure), and predict the bond order in the molecule. The Figure shows the electrons in the two atomic orbitals.

He_2 has four electrons, so two go into the bonding orbital and two into the antibonding orbital (Fig. 13, p. 19). The electron configuration of the molecule is $(\sigma 1s)^2 (\sigma^*1s)^2$. Molecular orbital theory predicts that there are zero net bonding electrons, so the bond order is zero and the molecule should be unstable. Certainly He_2 has never been observed in its ground state. This negative result is another success of the theory, and the prediction contrasts with the more empirically based theories which account for the non-existence of He_2 by the stability of the electronic configuration $s^2 p^6$.

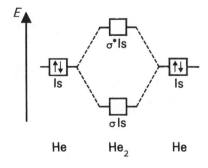

Figure 12   Energy level diagram for He_2

8.3.1 Other orbital combinations

Until now we have considered the combination of s orbitals only. You may be wondering whether the combination of other orbitals can generate molecular orbitals. This is equivalent to asking whether two hydrogen atoms can combine if their electrons are excited to higher energy orbitals.

> What condition is necessary for a suitable bonding orbital to result from a linear combination procedure?

For our purposes, it is sufficient to consider only s and p orbitals. The combination of two 1s orbitals has already been dealt with. The case of two 2s orbitals is completely analogous, two molecular orbitals, $\sigma 2s$ and $\sigma^* 2s$, are produced. There are two other cases to consider: s with p, and p with p. The p orbitals have directional properties (the value of ψ depends on the direction), so it is necessary to consider how the two atoms approach each other. Suppose that this approach is along the x axis. Figure 14 shows the combinations $s + p_x$ and $s + p_z$ (the combination $s + p_y$ is equivalent to $s + p_z$ but is less easily represented in the xz plane of the paper).

The two atomic orbitals must overlap to give an increase in electron density between the nuclei.

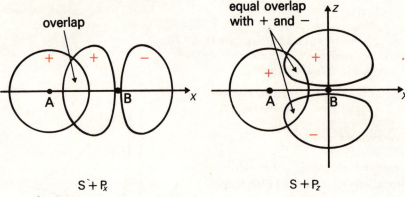

$$S + P_x \qquad\qquad S + P_z$$

Figure 14   Combination of s and p orbitals.

> What can you say about the stability and symmetry of these two linear combinations. Remember that the molecular wave function is given by
>
> $$\psi = N(\psi_A + \psi_B)$$
>
> and that the p_x orbital is symmetrical with respect to rotation about the x axis, but that the p_z orbital is not.

The combinations $s + p_x$ generates a bonding molecular orbital with σ symmetry (and, of course, a corresponding antibonding orbital for the combination $s - p_x$). The condition of orbital overlap is satisfied and results in an increase of electron density between the nuclei. Because both parent orbitals are symmetrical about the x axis, the parent orbital must also be symmetrical and is therefore a σ orbital.

In the second combination, $s + p_z$, there is no resultant overlap of the orbitals, because the sign of the wave function is different in the two lobes of the p orbital and they overlap equally with the s orbital. No molecular orbital is produced by this combination, because the overlap in the positive and negative regions of the p orbital cancel out exactly.

The symmetry conditions demonstrated by these two examples are quite general. Only if the two atomic orbitals have the same symmetry about the line joining the two nuclei, is a molecular orbital formed.

Now let us consider the linear combination of p orbitals. Figure 15 shows three possible cases.

> Will molecular orbitals be formed from any of these combinations? If molecular orbitals are formed, do they have σ symmetry?

(a)   The two p_x orbitals overlap to give a σ orbital. The reasons for the σ symmetry are the same as given in the $s + p_x$ combination.

(b)   Similarly the $p_x + p_z$ combination (and also $p_x + p_y$) is analogous to $s + p_z$; no molecular orbital is generated because the positive and negative overlap cancel out.

(c)   A new feature is apparent in the combination $p_z + p_z$. Both atomic orbitals have a nodal plane ($\psi = 0$) in the xy plane. The electron density along the x axis of the molecular orbital, which is the internuclear axis, will therefore be zero. However the electron density above and below the xy plane and between the two nuclei will increase when the molecular orbital forms. This is shown by the overlapping regions of the atomic orbitals in Figure 15 c. Increase in electron density in these regions binds the nuclei together and this linear combination produces a bonding orbital.

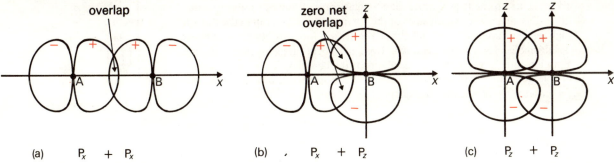

(a) $P_x + P_x$ (b) $P_x + P_z$ (c) $P_z + P_z$

Figure 15   Combination of two p orbitals.

Obviously this molecular orbital does not have σ symmetry. An orbital with a nodal plane through the internuclear axis is called a π *orbital* and has π symmetry. The combination $2p_z + 2p_z$ generates the $\pi 2p_z$ orbital. The combination $2p_y + 2p_y$ generates a molecular orbital with similar properties and the two orbitals are degenerate.

By examining the overlap and symmetries of the orbitals shown in Figure 15 we conclude that three of the combinations generate molecular orbitals. Like the combination of two 1s orbitals, antibonding combinations are also formed by combining wave functions of opposite sign:

$$\sigma^* 2p = 2p_x - 2p_x$$
$$\pi^* 2p_z = 2p_z - 2p_z$$
$$\pi^* 2p_y = 2p_y - 2p_y$$

In Unit 9 we shall see that $\pi 2p$ orbitals are important in the bonding in molecules like O_2 and N_2. In this Unit we concentrate on σ orbitals.

In this Section we have considered only the symmetry requirements for orbital formation. In the next Section we shall see that there are also energetic conditions.

Now summarize what you consider to be the main points of this Section and compare your summary with that given below.

Summary

1   Electrons in molecules are also represented by wave equations, although approximations are again necessary. One successful approximation involves adding the atomic wave functions; the linear combination of atomic orbitals.

2   The 1s orbitals of two similar atoms generate two molecular orbitals: a bonding orbital ($\sigma 1s$), in which electrons concentrate in the region between the nuclei and stabilize the molecule, and an antibonding orbital ($\sigma^* 1s$), in which electrons make the molecule unstable with respect to the atoms. H_2 is predicted to be stable and He_2 to be unstable.

3   Other orbital combinations can also generate bonding molecular orbitals when they overlap to increase the value of ψ in the internuclear region. For s and p orbitals, these combinations are: $s + p_x$ and $p_x + p_x$, giving σ orbitals; and $p_y + p_y$ and $p_z + p_z$ giving π orbitals. Corresponding antibonding orbitals are formed by combining atomic orbitals of opposite signs, e.g. $p_x - p_x$. No other combinations of s and p orbitals produce molecular orbitals.

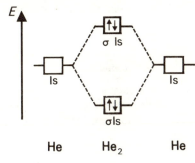

Figure 13   Energy level diagram for He_2.

SAQ 5 (Objective 7)   Two excited hydrogen atoms approach each other along the x axis. Which combinations of excited states 1 to 4 (right) are capable of producing a bonding orbital? Are the bonds formed σ or π bonds?

1   2s and $2p_x$
2   2s and $2p_z$
3   $2p_x$ and $2p_y$
4   $2p_y$ and $2p_y$

SAQ 6 (Objective 4)   Using the energy level diagram (Fig. 10) decide whether the ions H_2^+ and He_2^+ are stable in their ground states, and what the bond order in each ion is. You can apply the same method as that used for H_2 and He_2.

19

8.4 Bonding in the hydrides

In this Section we extend molecular orbital theory to diatomic molecules that are formed from dissimilar atoms, and to polyatomic molecules. Examples of these are found in the compounds that hydrogen forms with the elements of the second Period, lithium to fluorine.

The simplest hydride to begin with is lithium hydride. Throughout this Section we confine our interest to discrete molecules, e.g. the molecule LiH. Lithium hydride occurs normally as a white solid with an ionic structure similar to NaCl (Unit 2). Discrete gaseous molecules of the compound can be prepared although at high temperatures and with difficulty. The reason for restricting our interest to single molecules is clear enough. Even for such simple systems, drastic assumptions are necessary in the calculations; the complexities of systems containing large numbers of atoms are beyond the scope of present methods. We begin with the simplest cases, the diatomic molecules.

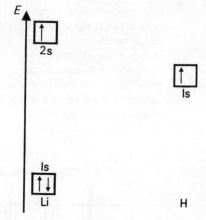

Figure 16   Energy levels in Li and H.

8.4.1   Diatomics: LiH and HF

Lithium hydride, LiH

The energy levels and orbital occupancy of Li and H atoms are shown in Figure 16. Two problems arise immediately. Which atomic orbitals do we choose to generate molecular orbitals, and how do we combine them?

Here we might be tempted to introduce an empirical factor into the argument. Lithium tends to form univalent compounds, LiH for example. This suggests that we should utilize only one lithium orbital, the 2s, since it is energetically more readily available. In fact, there are theoretical grounds for arriving at the same conclusion. In the linear combination the energy of a molecular orbital depends on the *difference* between the energies of the parent atomic orbitals. Atomic orbitals of very different energy do not generate stable molecular orbitals.

We can get some idea of the energies of electrons in the different orbitals from the values of the ionization energies. Although we would not be justified in comparing values of the successive ionization energies of lithium, because repulsive forces between electrons decrease as the atom is further ionized, these values do tell us that the electron in the 1s orbital is much more tightly held by the nucleus than is the electron in the 2s orbital (Table 1).

This energetic condition for molecular orbital formation imposes a general limitation on the combination of atomic orbitals; molecular orbitals are only formed from orbitals of the outer quantum shell (the valence shell). The 1s orbital of Li is energetically unfavourable (the 1s electrons are held too strongly by the Li atom), and the LiH problem involves essentially only two electrons and two atomic molecules.

The combination must be Li 2s plus H 1s, but, because these orbitals have different energies, the most favourable or lowest energy combination involves unequal contributions from the two wave functions.

$$\psi = N(\psi_{Li2s} + \lambda\psi_{H1s})$$

where λ is a factor which tells us the relative contributions of the Li 2s and H 1s atomic orbitals to the molecular orbital.

In the formation of H_2, $\lambda = 1$. Now we choose λ to give the most stable molecular orbital. Again, this type of optimization procedure is easily tackled with a computer.

The criterion for choosing a value for λ illustrates an important principle in molecular orbital theory. It should not be possible to calculate a set of wave functions which predict a molecule more stable than is found experimentally. Accordingly, we always select the lowest energy solution as the best approximation. Nature always selects the lowest energy solution possible, and if our theory is correct we should do the same.

As before the two atomic orbitals generate two molecular orbitals, one bonding and one antibonding. The two electrons occupy the bonding molecular orbital (Fig. 17), which is a σ orbital because both parent atomic orbitals are spherical.

Table 1. Ionization energies of hydrogen and lithium

I_1H 1 314 kJ mol^{-1}
I_1Li 519 kJ mol^{-1}
I_2Li 7 279 kJ mol^{-1}
I_3Li 11 811 kJ mol^{-1}

There are two bonding electrons, so the bond order is one.

Figure 6 shows that the hydrogen atom holds its electron more tightly than the lithium atom holds its 2s electron. The molecular orbital is a combination of the atomic orbitals and therefore retains this difference to some extent. This is reflected in the calculated valence electron distribution in the molecule, which shows clearly that the electrons tend to concentrate around the hydrogen atom; Figure 18 is a contour of valence electron density in LiH. Unequal charge distribution between the two atoms is called an electric dipole; the molecule has a dipole moment (Unit 1). Evidently the hydrogen atom is more effective than lithium in attracting electrons, a property that is termed electronegativity. We shall return to this concept later in the Unit.

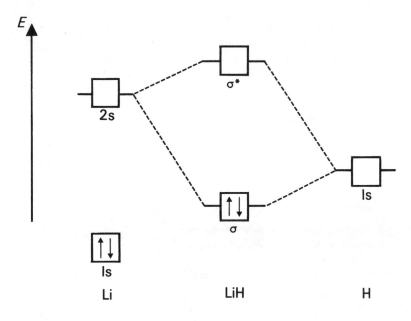

Figure 17   Orbital occupancy in LiH.

The unequal sharing of electrons can be regarded as a partial charge on the two atoms:

$$\overset{\delta+}{\text{Li}}\!-\!\overset{\delta-}{\text{H}}$$

The lithium atom, with less than a half share of the bonding electrons, is labelled $\delta+$. We say that the molecule has 'ionic character'.

In molecular orbital theory, we find the possible beginning of a unified explanation of covalent (e.g. H_2) and ionic bonding. (In Unit 7 we discussed G. N. Lewis's search for such an explanation.) If ionic and covalent bonds are merely extreme cases of electron sharing, molecular orbital theory may be able to accommodate both.

Hydrogen fluoride, HF

Hydrogen fluoride exists as a gaseous diatomic molecule. The electronic configuration of fluorine is $1s^2\,2s^2\,2p^5$.

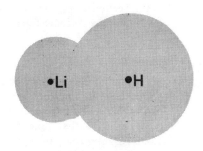

Figure 18   Distribution of valence electrons in LiH.

> What orbitals would you expect to be involved in bond formation?
> (Assume that the internuclear axis is the x axis.)

Linear combinations of F $2p_x$ and H $1s$ are expected. The combination is similar to the one shown in Figure 14 (s + p_x).

> Given the atomic ionization energies:
>
> $$I_1\,(\text{H}) = 1314\ \text{kJ mol}^{-1}$$
>
> $$I_1\,(\text{F}) = 1680\ \text{kJ mol}^{-1}$$
>
> what can you say about the electron distribution in the HF molecule? Refer to the LiH molecule if you are uncertain.

The fluorine atom holds its electron more tightly than hydrogen does. Because the molecular orbital is formed by combination of the atomic orbitals, it retains

21

some of the parent orbital characteristics. So we expect the fluorine atom to hold the bonding electrons more tightly than hydrogen. In other words we expect HF to have some ionic character which we can represent as

$$\overset{\delta+}{H}\!-\!\overset{\delta-}{F}$$

However, the difference in atomic ionization energies is much smaller than between H and Li, and the electric dipole of HF is correspondingly smaller than for LiH.

Molecular orbital theory can go one step further here and predict from the calculated electron distribution in the molecule a value for the dipole moment. The agreement with experiment is remarkable (Table 2).

Table 2   Dipole moments*/coulomb metre $\times$ 10^{-30}

	Experimental	Recent calculation
LiH	19.6	19.9
HF	5.8	6.5

* Dipole moments will be discussed in Unit 10.

8.4.2   Polyatomic hydrides—the shapes of molecules

Having examined the two diatomic hydrides that are formed by the elements of the second period, we now turn to the polyatomic hydrides. Immediately a new problem arises. Not only do we want to predict or confirm the stabilities of the molecules, we are now interested in their shapes. How does molecular orbital theory predict the shapes of molecules?

In Section 8.1 you saw that some atomic orbitals, e.g. 2p and 3d, have directional properties; they extend in certain directions from the nucleus. If these orbitals are involved in linear combinations with 1s orbitals of hydrogen atoms, molecular orbital theory will predict the bond angles in the compounds.

In this Section we examine the compounds that hydrogen forms with the elements Be to O in the Periodic Table.

Beryllium hydride

Beryllium has the electronic configuration $1s^2 2s^2$. Two atomic orbitals are fully occupied, so it is not obvious how beryllium forms bonds to hydrogen. With two valence shell electrons ($2s^2$) beryllium might be expected to form a compound BeH_2, in which the 1s atomic orbitals in Be retain their identity with the Be atom. However the formation of BeH_2 would involve the two 2s electrons, and it is not clear what the shape of the molecule would be.

In fact BeH_2 has never been observed as discrete molecules and the compound exists only as a solid polymer $(BeH_2)_n$. We shall discuss further the possibility of the existence of BeH_2 in Unit 9.

Boron hydride

With the electronic configuration $1s^2 2s^2 2p$, boron might be expected to form a compound BH_3 (three valence shell electrons), although again the shape predicted for the compound is not obvious. The compound BH_3 has not been observed, and boron forms a series of compounds with unusual stoichiometries, e.g. B_2H_6, which we discuss in Unit 9.

Methane, CH_4

Carbon forms a very large number of compounds with hydrogen, some involving multiple bonds (S100, Unit 10). In the singly bonded compounds, the carbon atom always has four bonds arranged tetrahedrally, and in the simplest compound, methane, these four bonds are equivalent. The theory of valence shell repulsion provides a satisfying explanation of this geometry.

What is the electronic configuration of carbon?

In its ground state it has the configuration $1s^2 2s^2 2p_x 2p_y$ with two unpaired electrons, If we accept the concept of the electron pair bond, we expect carbon to be divalent. Compounds formed with hydrogen, using the two 2p orbitals of carbon and 1s orbital of hydrogen, would have two bonds at right angles to each other. The compound methylene CH_2 is known, although it reacts vigorously with hydrogen to produce methane. The formation of four bonds (from four unpaired electrons) requires an excitation of some sort. The transfer of a 2s electron in carbon to the $2p_z$ orbital is energetically easiest. This gives the electronic configuration $1s^2 2s^2 2p_x 2p_y 2p_z$. However, the process requires energy, 6.7×10^{-19} J per atom (about 400 kJ mol^{-1}).

It is still not easy to see how the atom can form four equivalent bonds; the p orbitals have directional properties; the s orbital is spherical; and s and p orbitals have different energies.

A solution to this problem was proposed by the American scientist, Pauling, in 1931.

To begin with we must accept the notion that the excitation $2s \rightarrow 2p$ is possible to give the carbon atom the electronic configuration $2s2p^3$. You may object to this procedure because an excitation requires energy. At this stage such an objection is valid, and we shall return to this problem shortly. Meanwhile, if we accept that there are four orbitals, 2s, $2p_x$, $2p_y$ and $2p_z$, each occupied by a single electron, we are still faced with the problem of the shape of methane. Pauling's solution to this problem was to combine the four wave functions which describe these four orbitals in such a way as to produce four different but equivalent atomic orbitals. This combination of s and p orbitals is called *hybridization*, and the resultant hybrid orbitals have directional properties. If the four sp^3 hybrid orbitals are equivalent, they must be directed tetrahedrally from the carbon atom. When we represent the orbitals with contour surfaces, each consists of two lobes (Fig. 19). Usually we are concerned only with the larger lobes which are directed tetrahedrally and these are often shown elongated for convenience, as in Figure 20.

hybridization

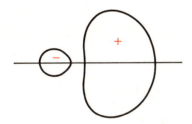

Figure 19 sp^3 hybrid orbital.

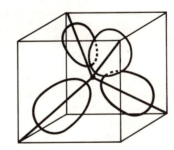

Figure 20 sp^3 hybrid orbitals.

In the formation of molecular orbitals in the molecules we have discussed so far, the atoms are arranged so that the atomic orbitals overlap (the linear combination procedure). With the sp^3 hybrid orbitals, it is easy to see that this overlap of the H 1s orbitals with the C sp^3 orbitals is maximized when the hydrogen atoms are arranged tetrahedrally around the carbon. Linear combination of one sp^3 with one 1s orbital of hydrogen is therefore the next step.

Does the molecular orbital produced have σ symmetry?

The sp^3 hybrid orbital is symmetrical about the tetrahedral axis (Fig. 19), which is also the interatomic axis, so the molecular orbital is also symmetrical about this axis. Figure 20 shows one of the advantages of the hybridization approach. The four molecular orbitals are located along the internuclear axes and we can identify them with the four C—H bonds. In this way the ability of carbon to form four bonds offsets the energy requirement of the excitation $2s \rightarrow 2p$.

Thus, by introducing the idea of hybrid orbitals, we are able to account for the shape of the methane molecule and the fact that carbon is tetravalent. It is important to recognize that the sequence of steps that we take in no way represents what happens when a carbon atom combines with four hydrogen atoms to form methane. The method is not meant to represent any physical process; it is merely a pathway to understanding the bonding in methane.

The idea of hybridization may seem inconsistent with our previous descriptions of chemical bonding. In fact, molecular orbital theory offers a different approach to this problem In the theory we used for diatomic molecules we were concerned with the formation of molecular orbitals from atomic orbitals that complied with the energy and overlap conditions.

According to the energy requirement we should consider all the valence shell orbitals: 2s, $2p_x$, $2p_y$ and $2p_z$ of carbon and 1s of the four hydrogens. Linear combination of these eight atomic orbitals produces eight molecular orbitals, four bonding and four anti-bonding. The eight valence shell electrons (four from carbon and four from the four hydrogens) occupy the lowest energy orbitals according to Pauli's principle, two in each orbital.

It turns out that the most stable arrangement occurs when the hydrogen atoms are arranged tetrahedrally. However, although there are four occupied bonding orbitals in the molecular orbital description of CH_4, these orbitals are not localized along the specific internuclear axis as they are in the hybridization model. Thus the molecular orbital description gains in being less contrived, but suffers in that we can no longer identify a particular orbital with a chemical bond.

We shall now examine how the theories described so far account for the bonding in NH_3 and H_2O. The results are similar for the two; for simplicity we shall examine H_2O first.

Water, H_2O

The simplest hydride of oxygen is water, H_2O, which exists as stable discrete molecules in the gas phase. The molecule has two equivalent bonds and the bond angle is found to be 104.5°.

You might recognize this angle as being close to the tetrahedral angle and so be tempted to say that the oxygen atom is sp^3 hybridized with two of the sp^3 orbitals in linear combination with the H 1s orbitals. Oxygen has the electronic configuration $1s^2 2s^2 2p^4$; there are six valence shell electrons. Paired with the two 1s electrons of the hydrogens, two of these valence electrons can occupy the molecular orbitals formed by linear combination. This leaves four electrons to occupy the remaining two hybrid orbitals. These two pairs of electrons correspond to the lone pairs of electrons.

This picture seems reasonably satisfactory for H_2O and agrees with the prediction of valence shell repulsion theory; the tetrahedral bond angle is within 5° of the observed bond angle. But we find that this approach works less well when we examine the hydrides of other elements with the s^2p^4 electronic configuration. Their bond angles are considerably smaller than 109° (right).

Although sp^3 hybridization is often used to describe the bonding in H_2O, molecular orbital theory offers a simpler approach which leads to slightly different predictions. In its valence shell oxygen has two atomic orbitals fully occupied. Because the three 2p orbitals are equivalent (degenerate) we can single out any one of these. Let us say that the 2s and $2p_z$ orbitals are occupied. Suppose that the remaining two, $2p_x$ and $2p_y$, are used in linear combination with the two 1s orbitals of the hydrogen atoms. This requires that the hydrogen atoms are placed along the x and y axes for maximum overlap (Fig. 21). These combinations yield molecular orbitals which are localized along the internuclear axes.

Do the molecular orbitals have σ symmetry, and what bond angle is predicted?

Both H 1s and O $2p_x$ orbitals are symmetrical with respect to rotation about the x axis, so this combination generates a σ orbital (compare with Fig. 14a). An exactly similar argument applies to the other molecular orbital. Figure 21 suggests that the most stable arrangement (maximum overlap of atomic orbitals) occurs at a bond angle of 90°.

Bond angles

H_2S	92.0°
H_2Se	91°
H_2Te	89°

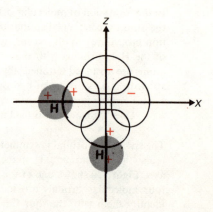

Figure 21   Orbital overlap in H_2O.

In the theories of hybridization and valence shell repulsion, two electron pairs, the lone pairs, are not involved in bonding. The molecular orbital equivalent of these lone pairs is the 2s and $2p_z$ orbitals, which contain two pairs of electrons which do not contribute to the bonding. We call these orbitals *non-bonding orbitals*.

Evidently the hybridization model comes closer to the observed bond angle in H_2O, but for the analogous compounds in higher periods the 90° prediction of simple molecular orbital theory based on the geometry of 2p orbitals is more accurate.

Both theories are approximations, although the simple molecular orbital theory has the advantage that it is based on the shape of the atomic orbitals and appears less contrived. More sophisticated calculations use the molecular orbital approach, but attempt to determine how the energy of the molecule varies with the bond angle. The lowest energy corresponds to the most stable molecule, and a recent calculation gave a bond angle of 106°.

Ammonia, NH_3

The simplest hydride of nitrogen, NH_3, has three equivalent bonds in a trigonal pyramid arrangement with a bond angle of 106.7° and a bond length of 101.2 pm. Nitrogen has the configuration $1s^2 2s^2 2p_x 2p_y 2p_z$. If a simple linear combination is made of the three 2p orbitals with the 1s orbitals of hydrogen, we expect three molecular orbitals directed along each of the three perpendicular axes, x, y and z. The shape of the molecule is therefore predicted to be a trigonal pyramid with bond angles of 90° (the angle between the 2p orbitals).

The predictions of valence shell repulsion theory are rather better. NH_3 contains eight valence electrons which pair to give a tetrahedral arrangement (bond angle 109.5°). One electron pair is not involved in bonding, the lone pair. The molecular orbital equivalent of the lone pair is the 2s orbital.

If hybridization occurs, the bond angles are the same as in methane, 109.5°, in agreement with the prediction of valence shell repulsion theory.

Again, more sophisticated calculations involve the determination of the total energy of the molecule as a function of the bond angles.

One recent calculation has achieved extremely accurate estimates of molecular properties: bond angle 107.2°, bond length 100 pm. At present, calculations of this accuracy can be made only for fairly simple molecules.

The same calculation gives an energy of formation from gaseous atoms within about 30 per cent of the experimental value. Contrast this with the accuracy of calculating the bond energy of H_2.

> Can you see why it is difficult to calculate bond energies and energies of formation of larger molecules? If not, look at Figure 6.

These calculations involve the determination of the energies of the valence electrons, but the energies of the electrons in the inner (1s) orbital are very much greater. In fact, the *total* electron energy of NH_3 has been determined to better than 0.01 per cent of the experimental value.

In this Section we have considered compounds which have certain properties:

1 they form discrete molecules in the gas phase;

2 they contain only one complex atom (and hydrogen atoms);

3 they contain only a small number (not more than five) atoms.

Most chemicals are not as simple as this, and generally do not even exist as discrete molecules.

Although molecular orbital theory offers satisfactory results for simple compounds, it does not cope well with polyatomic molecules that do not contain hydrogen. We still have a long way to go to a molecular orbital description of a simple solid compound like $AlCl_3$ and an estimate of its ionic character.

Now summarize what you consider to be the main points of this Section and compare your summary with that given overleaf.

Summary

1   Molecular orbital theory describes the bonding in those hydrides of the second row elements that form discrete molecules.

2   Molecular orbital formation is subject to an energy criterion: the parent atomic orbitals must have similar energies.

3   The orbital descriptions of LiH and HF account for the dipole moments in the molecules.

4   In methane the formation of four equivalent, tetrahedrally disposed bonds can be described in terms of sp^3 hybridization.

5   Simple orbital overlap predicts bond angles of $90°$ in NH_3 and H_2O. The observed bond angles are rather closer to the sp^3 hybridization picture.

SAQ 7 (Objective 6)   The bond angle in the molecule PH_3 is found to be $93°$. Compare this value with the predictions of electron pair repulsion and molecular orbital theories.

SAQ 8 (Objective 1)   What is the role of the 3s electrons of P in the PH_3 molecule according to the molecular orbital theory?

SAQ 9 (Objectives 1 and 6)   The valence electron configuration of Si is $3s^2 3p_x 3p_y$. The molecule silane SiH_4 has a tetrahedral structure. How is this explained in molecular orbital theory?

Study comment

We will return to the hydrides in Section 8.8.

In the meantime we are going to examine the basis in atomic structure of the Periodic Law. This study forms a bridge between earlier Sections of this Course which were mainly about metals, and later Sections which will be mainly about non-metals. In the final Sections of this Unit we begin the descriptive chemistry of the non-metals, with a study of hydrogen. In Unit 9, we continue both the bonding theory, and the descriptive chemistry, with a study of the elements of the second row.

8.5   Metals, non-metals, and the Periodic Law

Think back to how the Periodic Table was built up, in Units 7 and 8 of S100.

Can you write down in one sentence a simple statement of the Periodic Law?

Perhaps you wrote, that the chemical properties of the elements arranged in order of atomic number (Z) vary periodically, that is, that characteristic patterns of behaviour, such as that of the alkali metals, recur at particular intervals.

From S100, Unit 6 we know that the electronic structures of the atoms arranged in order of atomic number vary periodically. It seems likely that the chemical properties are periodic because of their dependence on electronic structure.

Periodic Law due to shell structure

In the next three Sections we shall see how the shell structure of the atom leads to the pattern of metals and non-metals that we observe in the Periodic Table. For the moment we confine ourselves to the typical or main-group elements, since the transition metals form exceptions to some of our generalizations.

As we have seen in earlier Units, metals are elements which lose electrons more readily than non-metals. They lose them to suitable acceptors, such as a halogen, or to the pool of electrons held in common in a metallic structure.

As Figure 22 shows, the rows of the Periodic Table progress from metals on the left to non-metals on the right.

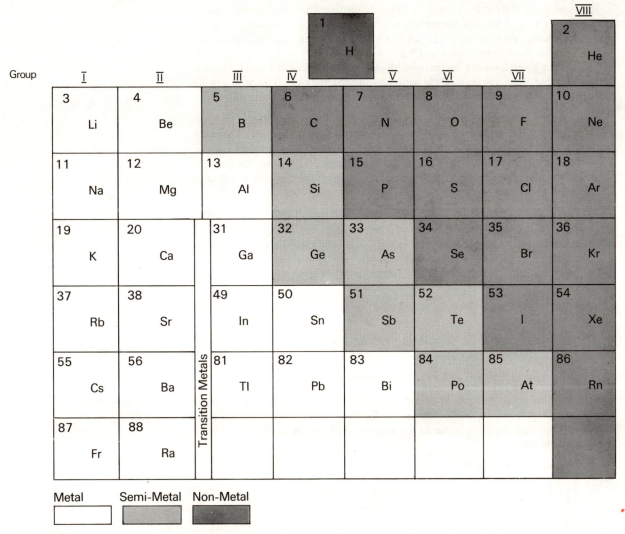

Group	I	II		III	IV		V	VI	VII	VIII
						1 H				2 He
	3 Li	4 Be		5 B	6 C	7 N		8 O	9 F	10 Ne
	11 Na	12 Mg		13 Al	14 Si	15 P		16 S	17 Cl	18 Ar
	19 K	20 Ca		31 Ga	32 Ge	33 As		34 Se	35 Br	36 Kr
	37 Rb	38 Sr	Transition Metals	49 In	50 Sn	51 Sb		52 Te	53 I	54 Xe
	55 Cs	56 Ba		81 Tl	82 Pb	83 Bi		84 Po	85 At	86 Rn
	87 Fr	88 Ra								

Metal Semi-Metal Non-Metal

We can follow this progression in the physical and the chemical properties of the elements. Thus, typical metals (contrasted with typical non-metals) are lustrous, conduct electricity, and so on. As to the chemical progressions, we have seen some examples, in Units 5, 6 and 7.

Figure 22   Metals, semi-metals and non-metals.

> Can you list some of these?

Across the row, the acid character of oxides (of the same formula type) increases. Hydroxy compounds decrease in basic character, and increase in acid character (Unit 6, Section 6.2.3).

chemical periodicity

Chloride structures change from ionic to covalent across the third row (Unit 7, Section 7.4), and this is true of the other rows.

Figure 22 shows also that metallic character tends to increase down the Groups. Groups III to VI proceed from a non-metal at the top to a metal at the bottom.

In Group VII the non-metals at the bottom (iodine and astatine) have some semi-metal properties (see Units 10 and 11). In Group II the metal at the top (beryllium) forms covalent rather than ionic bonds, and lithium in Group I shows a slight tendency in this direction, compared with the other alkali metals. (The position of lithium at the head of the Electrochemical Series of the metals, in order of their readiness to be oxidized by aqueous H^+, is discussed in Unit 9; and chemical progressions in the Groups are discussed in Section 8.7 of this Unit.)

Notice that these changes from metal to non-metal in the Periodic Table are gradual; the elements in the middle have semi-metal or metalloid properties (see Unit 10). For example, although graphite, black phosphorus, grey selenium, and iodine are classified as non-metals, they all have some metallic lustre, and some conduct electricity.

Notice, too, that the changes in the Periodic Table are not smooth. Graphite is an electrical conductor and diamond an insulator, although both are pure carbon. Thus the physical property depends on the structure, and non-metal structures vary. Chemical properties, too, depend on the structure; ozone is more reactive than oxygen, and white phosphorus more reactive than black phosphorus. Reactivity sequences (as given in Unit 5) vary with the reagent, for although a given factor in the reaction (the lattice energy of a salt that is formed, for example) may vary smoothly in periodic sequence, the overall reaction usually depends on several factors, each with its own form of periodic variation.

But the overall periodic trends remain. In order to understand them, we shall look now at the periodicity of 'atomic' properties of the elements. These are related, on the one hand, to the electronic structure that underlies the rows and Groups of the Periodic Table (cf. Unit 7 of S100) and, on the other hand, to the 'metallic character' of the elements, that is, to the properties that distinguish metals from non-metals.

Such properties are the *ionization energy* and the *electron affinity* which is useful in describing non-metals. The ability of a bound atom to attract electrons is its *electronegativity*. Another property we shall find important is the atomic size, which may be expressed as a *covalent radius* (r_c), or an *ionic radius* (r_i); this we explore also in the TV programme for this Unit. Values for all these quantities are given in your *Data Book**.

8.6 Periodicity of atomic properties of the elements

8.6.1 The increase in the ionization energy across the row

The first ionization energies of the typical elements (see the *Data Book*, Section 9.1) are plotted against atomic number in Figure 23. Evidently, across the row, there is a large overall increase in the strength with which the atom holds its valence electrons.

ionization energy

The increase is rather a jerky one.

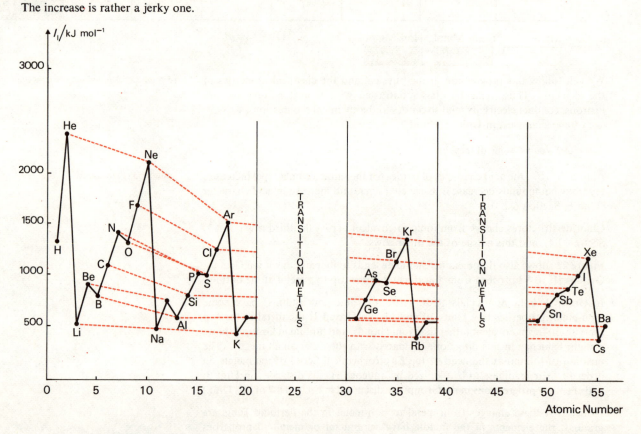

What is the physical significance of the maxima at the noble gas elements; and of the minima at the alkali metals?

Figure 23 First ionization energies of the typical elements

* The Open University (1973) S24-/S25- *The Open University Chemistry Data Book*, The Open University Press.

At these maxima, a very high energy is required to remove an electron, which means that the electronic configuration ($s^2 p^6$) is very stable. At the minima, an electron is more readily lost, to give a stable $s^2 p^6$ ion. This is familiar to us, as 'the stability of the noble gas electronic configuration', the octet.

Similarly, the local maxima show us that the s^2 and $s^2 p^3$ configurations are slightly more stable (and the local minima show that the $s^2 p$ and $s^2 p^4$ configurations are slightly less stable) than we would expect if the ionization energy increased steadily across the row. We explore these irregularities in Appendix 4 (Black).

With these small reservations, then, the trend in ionization energies is in accord with the decrease in metallic character across the row.

The ionization energy does not, however, give us direct information about the ability of an atom to add an electron to form an anion. For this we need the electron affinity.

8.6.2 The increase in electron affinity across the row

The electron affinity (*EA*) of an element is defined as the energy given out when the gaseous atom takes up an electron. Values are given in your *Data Book*, Section 10, and are plotted in Figure 24. Since the electron affinity is given by the ionization energy of the mono-anion, the values are known accurately for the non-metals that form anions, but they are more difficult to measure for the other elements.

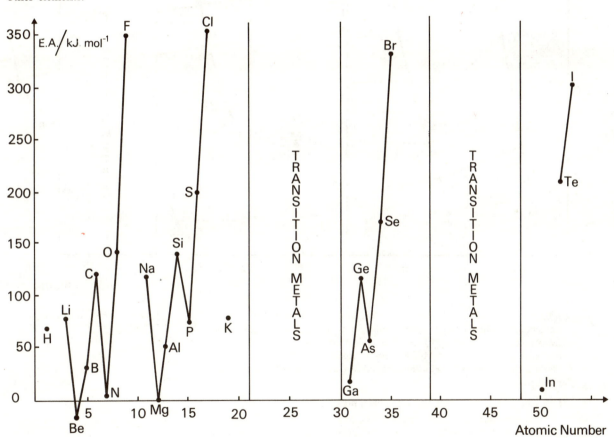

Figure 24   Electron affinities of the typical elements.

The electron affinity is always negative for the addition of a second electron, since this is repelled by the mono-anion. Thus for oxygen:

$$O(g) + e = O^-(g) \qquad EA = 147 \, kJ \, mol^{-1}$$

$$O^-(g) + e = O^{2-}(g) \qquad EA = -843 \, kJ \, mol^{-1}$$

Do you expect the electron affinity values to reflect the relative stabilities of the electronic configurations, as the ionization energies do? Do they?

Figure 24 shows that they do. The value for lithium ($s^1 \rightarrow s^2$) is relatively high, as is the value for carbon ($s^2 p^2 \rightarrow s^2 p^3$). These reflect the relative stability of the

29

s^2 and s^2p^3 configurations. The peaks are at the halogens, which form the closed shell anions, s^2p^6. Again, the low values for beryllium and for nitrogen tell the same story as the ionization energies do (Appendix 4).

8.6.3 The increase in electronegativity across the row

Figures 25 and 26 show the periodic variation of the Pauling and Mulliken electronegativities of the elements, which were mentioned in Unit 5, Appendix 1. Mulliken suggested in 1934–5 a scale of electronegativity (ϵ) defined as proportional to the arithmetic mean of the first ionization energy I_1 and the electron affinity (EA) of an atom, since each of these quantities is a measure of how strongly valence electrons are held by the atom: $\epsilon = \frac{1}{2}(I_1 + EA)$.

Mulliken electronegativity

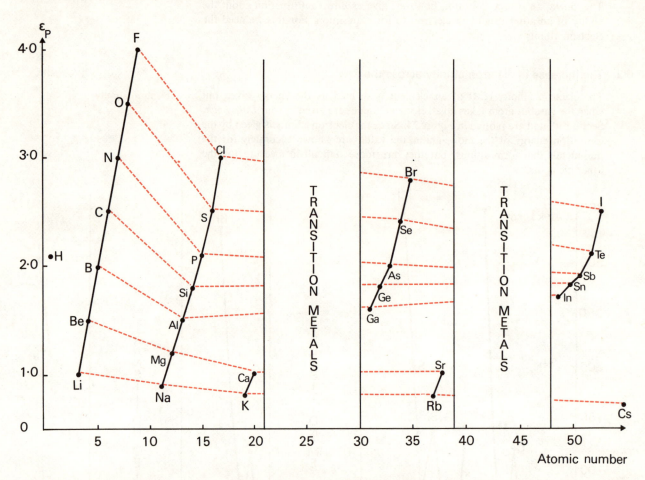

Figure 25 Pauling electronegativity of the typical elements.

SAQ 10 (Objectives 8, 10) Work out the Mulliken electronegativities of the elements from lithium to fluorine, using the I_1 and EA values given in your *Data Book*. If the mean in kJ mol^{-1} is multiplied by 4×10^{-3}/kJ mol^{-1}, this multiplying factor gives the Mulliken electronegativity of fluorine as 4, as for the Pauling electronegativity. Plot these Mulliken electronegativities against atomic number.

Can you see any difficulties in the Mulliken proposal?

One difficulty is that electron affinities are difficult to measure, and are not known for all elements. Another is that the values reflect properties of the free gaseous atom, rather than of the atom in a molecule*. Further, the definition refers to the loss or gain of only one electron, so that it applies less well perhaps to multivalent than to univalent atoms. The Mulliken electronegativity has the advantage that it is measured directly, but it has not the practical value of the Pauling electronegativity or the Allred–Rochow scale described below.

* In the free atom one can say that an electron goes into, or comes from, a p rather than an s orbital, say. In the atom in a molecule, as we saw in Section 8.4.2, electrons usually have a mixture of s and p character.

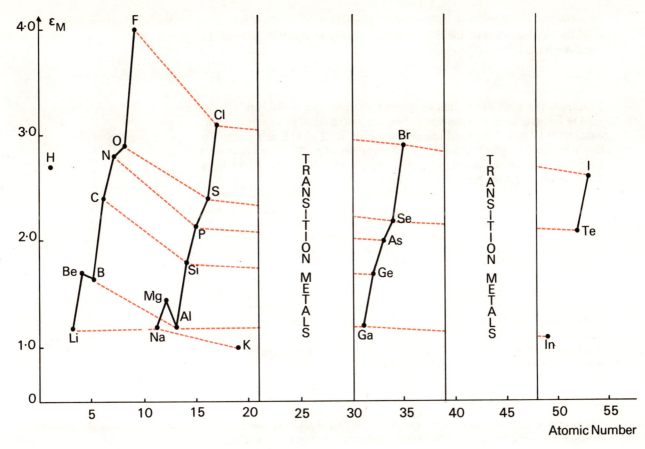

Figure 26 *Mulliken electronegativity of the typical elements.*

Pauling electronegativity

Electronegativity is most usefully defined as the power of an atom of an element to attract bonding electrons towards itself, in compounds with other elements (cf. Section 8.4.1). Thus, the separation of charge in a covalent bond A—B which has some polarity $\overset{\delta-}{A}$ — $\overset{\delta+}{B}$ depends on the difference in the electronegativities of A and B.

Pauling has described such bonds in terms of resonance (Unit 7, Section 7.5) between two hypothetical forms, one fully ionic A^-B^+, and the other non-polar A—B. The percentage ionic character is then estimated from the bond energies; it is the greater, the more stable the ionic relative to the non-polar form.

In 1932, Pauling proposed an electronegativity scale in terms of observed bond strengths, and this has the merit of being appropriate to atoms in molecules. The crux of the matter is, that the *energy term** E_{AB}, of a single bond between atoms A and B, is usually greater than the arithmetic mean of the values E_{AA} and E_{BB} for the single bonds A—A and B—B**. The difference, Pauling called the excess energy, or delta (Δ), where

$$\Delta_{AB} = E_{AB} - \tfrac{1}{2}(E_{AA} + E_{BB})$$

What has this extra energy to do with electronegativity?

In a polar bond $\overset{\delta-}{A}$—$\overset{\delta+}{B}$ extra energy can be attributed to the electrostatic attraction of the unlike charges.

Pauling found that $\sqrt{\Delta}$ is reasonably additive, so long as the atoms A and B have their normal valencies (we shall return to this point later). By additivity

---

* This is defined in Appendix 2 (White), and in your *Data Book*, Section 8.

** For bond energy terms of elements such as nitrogen, or sulphur, which do not form singly bonded diatomic molecules, Pauling used the energy terms for the singly bonded polyatomic molecules H_2N—NH_2 (hydrazine), HO—OH (hydrogen peroxide), S_8 (ordinary sulphur); and so on. Pauling has also suggested that the geometric mean be used instead of the arithmetic mean. Each has certain advantages under different circumstances.

31

we mean that it is possible to write down a consistent set of electronegativities, ϵ, of the elements, so that the difference $(\epsilon_A - \epsilon_B)$ gives the $\sqrt{\Delta}$ value of the bond A—B:

$$(\epsilon_A - \epsilon_B) = \sqrt{\frac{\Delta}{K}}$$

Pauling electronegativity

Frequently the additivity is sufficiently good for electronegativity values to be used to estimate bond energies where these are unknown. K is a scaling factor, equal to 96 kJ mol^{-1} for bond energies measured in kJ mol^{-1}, and a constant factor is added to give fluorine an electronegativity of 4. This is a convenient scale, in that the values for the second row elements increase in steps of 0.5, from 1 for lithium to 1.5 for beryllium, and so on.

> SAQ 11 (Objective 8)   Using the bond energy terms given in your *Data Book* (Section 8), determine the extra energy Δ in the Si—O and in the C—O bonds. From these, estimate the difference in electronegativity between silicon and oxygen, and between carbon and oxygen, and compare these differences with those given in Section 11 of the *Data Book*.

The assumption on which Pauling's method rests is not necessarily valid. This is that the extra energy in a bond is wholly due to its ionic character. Extra energy may in fact be due to partial multiple bonding, as in compounds of non-metals of the third row (see Unit 10). But we have an independent method of checking Pauling's values for the electronegativity of atoms in molecules, the one proposed by Allred and Rochow in 1958.

Allred–Rochow electronegativity

Allred and Rochow considered the two nuclei A and B and the electron pair between them (in a single bond AB) as a simple electrostatic system. The Coulomb law (S100, Unit 4, Section 4.2.2) then gives the force of attraction (F) between the nucleus A, and a valence electron shared with B, as

Allred–Rochow electronegativity

$$F = -\frac{kZ_{eff} e^2}{r_c^2} \qquad (1)$$

where k is a constant

$-e$ is the charge on an electron

r_c is the covalent radius (of atom A) which we define in the next Section (the values are in your *Data Book*, Section 12). We can ignore the sign of F, since we are not likely to forget the direction of the force.

$Z_{eff}\, e$ is the effective nuclear charge of the atom A.

What is the actual nuclear charge of atom A, with atomic number Z?

The nuclear charge of atom A is Ze.

The *effective nuclear charge* is the portion of the positive charge of the nucleus A that the valence electron actually 'sees', allowing for the other electrons in between (in the inner shells, for example). $Z_{eff}\, e$ is defined as Ze less a *shielding* or *screening* term; the other electrons on atom A are said to shield or screen some of the nuclear charge from the valence electron, when they are between it and the nucleus. Z_{eff} is then the effective atomic number, and is often used as a shorthand for the effective nuclear charge. It is of course the effective nuclear charge in units of e.

effective nuclear charge

electron shielding

The Z_{eff} approximation is explained in some detail in Appendix 2 (White). It is a device, much used in theoretical chemistry, whereby we can consider our atom A as 'hydrogen-like', consisting of the electron we are considering (here the valence electron) plus a *core* composed of the nucleus and the other electrons. We shall find Z_{eff} very helpful in understanding the make-up of the Periodic Table. Appendix 5 (Black) shows how Z_{eff} can be obtained from the ionization energy, for hydrogen and the alkali metals, and how it is estimated for more complicated atoms. Values of Z_{eff} are given in Table 3 (on p. 37), and we discuss its variation in the Periodic Table in Section 8.6.2.

core

In equation 1 above, the only variables are Z_{eff} and r_c. When Allred and Rochow plotted Z_{eff}/r^2 against the Pauling electronegativity ϵ_P for the elements, they found that they could draw a straight line that fitted the points quite well.

What does this tell us about the two electronegativity scales?

The linear relationship of the two sets of values, the one based on 'excess energy' and the other on an electrostatic model using Z_{eff}, supports our belief (or hope) that both are measuring the same thing, 'the ability of an atom in a molecule to attract the bonding electrons'.

As with the Mulliken electronegativity, the plot against ϵ_P is used to set up the Allred–Rochow scale. This gives the Allred–Rochow electronegativity as

Allred–Rochow electronegativity

$$\epsilon_{AR} = \frac{3\,590\,Z_{eff}}{r_c^2} + 0.74, \text{ for } r_c \text{ in pm.}$$

SAQ 12 (Objectives 8–10)   What is the Allred–Rochow value for the electronegativity of chlorine? How does it compare with the Pauling electronegativity?

Allred–Rochow electronegativity values are particularly useful when Pauling values are unobtainable, as for the noble gases krypton and xenon, which form bonds with fluorine or oxygen (see Unit 11) but not with themselves. ϵ_{AR} is 3.0 for krypton, and 2.8 for xenon. If you plot these values in Figure 25 you will see how well they fit.

The uses of electronegativity

Electronegativity arguments are useful in organic chemistry. The C—H bond in a hydrocarbon such as ethane, H_3C—CH_3, is non-polar, carbon and hydrogen having rather similar electronegativities; but if an electronegative atom, such as oxygen, or a halogen is attached to carbon, carbon develops a small positive charge and becomes more electronegative itself. Thus, reactions in which a polar reagent attacks carbon in an organic molecule are sensitive to the electronegativity of substituents.

The polarization of bonding electrons by a more (or less) electronegative substituent is called an *inductive effect*; the classic example is the higher acid strength of chloroacetic acid compared with acetic acid CH_3COOH (see Section 15 of your *Data Book*).

inductive effect

$$Cl \longrightarrow CH_2 \longrightarrow COO^- H^+$$

The arrows show the inductive effect of the chlorine, relayed through the intervening atoms, which makes it easier for H^+ to separate from the acid grouping, to leave the chloroacetate anion. This anion is stabilized (relative to the undissociated acid) by the presence of chlorine, which carries part of the anionic charge.

Do you expect dichloroacetic acid to be a stronger acid than chloroacetic acid?

Check your answer with your Data Book.

Similar arguments can be used in inorganic chemistry. We noted in Unit 6, Section 6.2.3 that higher and lower oxides of the same element have different acidities.

Which is the more acidic? Is there a relation between oxidation number and electronegativity?

Higher oxides are more acidic than lower oxides of the same element. This can be attributed to the inductive effect of the attached oxygen, making the central atom the more electronegative, the more oxygens are attached. Thus antimony pentoxide Sb_2O_5 is acidic, while antimony trioxide Sb_2O_3 is amphoteric and forms antimony salts with acids. Another description of this, is that the Sb^{3+} cation is relatively stable (or may be stabilized) but the Sb^{5+} cation is not, because of its very high charge*.

Similarly, higher oxyacids $X(OH)_mO_n$ are usually more acidic than lower acids (with a smaller value of n). An example is sulphuric acid, $S(OH)_2O_2$, compared with sulphurous acid $S(OH)_2O$ (Unit 10 and Unit 11, Appendix 1).

* We can generalize this, and say that a semi-metal (see Unit 10) behaves more like a metal in its lower than in its higher valence states.

The acidities are given in your *Data Book*, Section 15. As with the chloroacetic acids, the complex anion is the more stable (relative to the undissociated acid) the more electronegative atoms are attached to the central atom and share the negative charge.

Electronegativity values are often used as if each element had a standard value. But this is not so. We have seen how the electronegativity of an atom depends on its substituents or ligands* in a particular compound. Similarly, although hydrogen in ethane is neither acidic nor hydridic, acetylene $HC{\equiv}CH$ forms acetylides, such as potassium acetylide, $2K^+(C{\equiv}C)^{2-}$, described in Unit 9. Carbon is more electronegative when multiply bonded than when singly bonded.

ligand

You may have noticed that the Pauling and Allred–Rochow definitions specify single bonds. They also specify that the atoms should be in their normal valence states. Attempts have been made to refine the concept of electronegativity, by listing different values for different bond types or oxidation numbers (e.g. 1.8 for Fe^{II} compared with 1.9 for Fe^{III}); but then simplicity and generality are lost, and the precision may be illusory.

Electronegativity arguments (as well as electronegativity values) are often misused. Thus properties of the halides are often correlated with decreasing electronegativity down the Group (Figure 25) when the straightforward correlation is in fact with the size of the halide ion (see Section 8.6.6, and Units 7 and 11). Thus electronegativity arguments correctly predict (as above) the relative strengths of oxyacids, in which $H^+(aq)$ separates from oxygen in each case. But the aqueous halogen acids HX become stronger down the Group despite the decreasing electronegativity. We examine the relevant energy relationships in Unit 11, and find that the size and electronegativity of the halogen determine the observed sequence in a more complex manner than would appear at first sight.

Thus when we use electronegativity arguments, we must at the same time use our chemical intelligence, in its dual sense of information and understanding.

8.6.4 The decrease in covalent radius across the row

Figures 27 and 28 give the relative sizes of atoms of the typical elements, as we show in this Unit's TV programme. Since it is difficult to delimit an isolated atom (Section 8.1), we define the *covalent radius* r_c as half the distance between two like atoms, singly bonded, as in the molecules of ethane $H_3C{-}CH_3$, or of chlorine.

covalent radius

The term 'radius', however, suggests a spherical atom, and we know from Section 8.3 that covalently bonded atoms are distorted in the bond direction.

We therefore distinguish between the covalent radius and the *non-bonded radius* or *van der Waals radius, r_w* (Fig. 29) using as example the structure of solid chlorine (Fig. 30), determined by X-ray crystallography at low temperature. If we bear this distinction in mind, we can use the convention which pictures the singly bonded atom as a sphere with radius r_c. This convention is implicit in Figure 27, and in the building of space filling models, as in your Home Experiment for Units 2 and 3. It is a convenient symbolism.

van der Waals radius

In the TV programme we explain that the lengths of single covalent bonds should be used in the estimation of covalent radii. This is because bond length varies with the type, e.g. ionic, metallic, or covalent (single, double or triple).

We then show that a particular single bond has a roughly constant length in different compounds, e.g. the $C{-}C$ bond in ethane or diamond, and the $C{-}O$ bond in methanol (H_3COH) or methyl ether (H_3COCH_3).

With certain elements, there are difficulties in choosing a representative single bond covalent radius. The radius of hydrogen is rather variable (see Section 8.8). The covalent radii of N, O, and F are larger in the 'standard' molecules

---

* A ligand (from the Latin verb *ligare*, to bind) is an atom or group that is bound to the atom one is considering. (In organic chemistry the word 'substituent' is used.)

H_2N—NH_2 (hydrazine), HO—OH (hydrogen peroxide), and F_2 than in their bonds to hydrogen and carbon; we examine this anomaly in detail in Unit 9. On the other hand, the covalent radii of Si, P, and S are smaller in some apparently single bonds than in their bonds to hydrogen or carbon (this we examine in Unit 10). Some cases, however, are unexpectedly simple. You may have been wondering how we find the covalent radius of an alkali metal, which normally forms ionic compounds. In fact, the vapours contain small concentrations of M_2 molecules such as Li_2 (Unit 9) which can be measured spectroscopically.

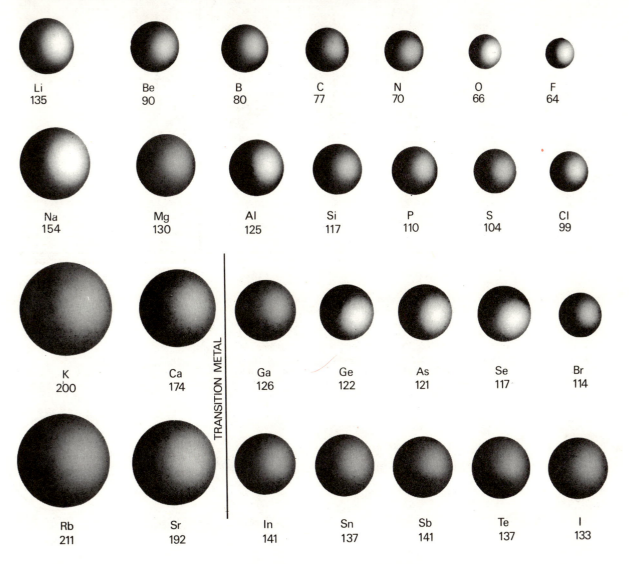

Figure 27   Covalent radii (r_c) of the typical elements.

Broadly speaking, then, covalent radii are additive, as are ionic radii (see Unit 7). We can choose for the elements a set of radii, r_c as in your *Data Book*, Section 12, which reproduce observed bond lengths, to within 10 per cent or better in practice. This is a useful measure of the relative sizes of atoms in molecules.

Figures 27 and 28 show that the sizes of atoms in molecules vary periodically. A remarkable feature of Figure 27, which we discuss in the TV programme, is that the atoms shrink markedly across the row of the Periodic Table, although they are getting heavier.

If, however, we reflect on the other trends in atomic properties that we have been discussing, i.e. the tendencies of the ionization energy, the electron affinity, and the electronegativity to increase across the row, we suspect that these are all telling the same story.

> Can you think of a unifying hypothesis to explain all these four trends in terms of atomic structure?

The hypothesis we develop is that the atoms are tending to hold their valence electrons more and more tightly, as we progress across the row; the nuclear charge is attracting the valence electrons more strongly.

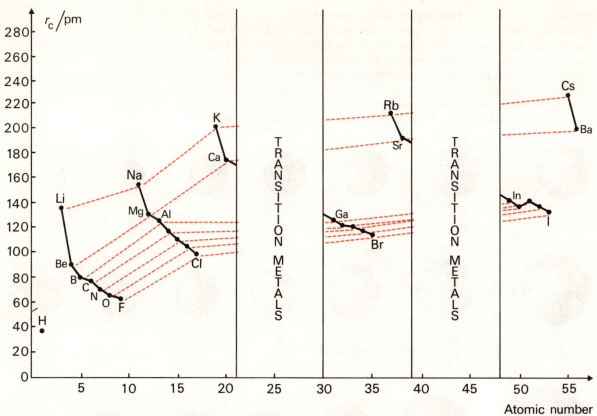

Figure 28 Covalent radii (r_c) of the typical elements. This plot is related to the one in S100, Unit 8, Appendix 3, which shows the periodicity of the atomic volumes of the elements. You may wish to consider why the two graphs differ. In some structures, such as those of the noble gases, the only internuclear distance is r_w. In others, such as those of the solid halogens, distances $2r_w$ and $2r_c$ are present. In O_2 and N_2 the bonds are multiple. Both r_w and r_c decrease across the row of the Periodic Table, but $r_w > r_c$ for a given atom. In addition the structure types change discontinuously across the row, for the lighter elements.

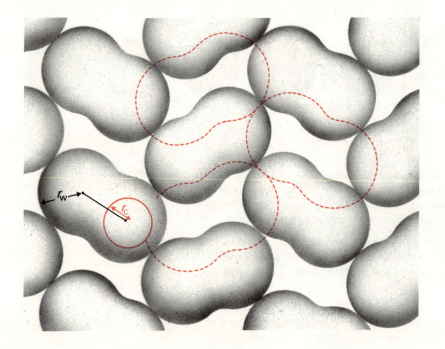

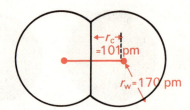

Figure 29 The chlorine molecule, showing the covalent and non-bonded radii.

Figure 30 Chlorine molecules in a layer of the crystal structure at −160 °C; the molecules drawn in red are in the next layer. In neighbouring layers, the molecules are staggered, to give a fairly close packing. The shortest internuclear distance between non-bonded atoms is 334 pm within a layer, compared with 370 pm between one layer and the next.

8.6.5 The effective nuclear charge and the changes across the row

We mentioned the idea of an effective nuclear charge in Section 8.6.3, when we were describing the Allred–Rochow electronegativity. *If you have not already read Appendix 2 (White), which explains the Z_{eff} approximation, read it now.*

Table 3 Z_{eff}, the effective atomic number

			H 1.0				He 1.7
Li 1.30	Be 1.95	B 2.60	C 3.25	N 3.90	O 4.55	F 5.20	Ne 5.85
Na 2.20	Mg 2.85	Al 3.5	Si 4.15	P 4.8	S 5.45	Cl 6.1	Ar 6.75
K 2.20	Ca 2.85	Ga 5.0	Ge 5.65	As 6.30	Se 6.95	Br 7.60	Kr 8.25
Rb 2.20	Sr 2.85						

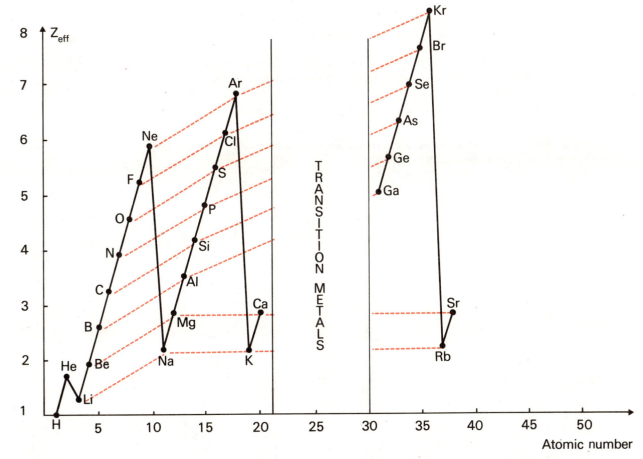

Figure 31 *Effective atomic number of the typical elements (as 'seen' by a valence electron).*

Appendix 5 (Black) describes how Z_{eff} values are estimated; the Slater values are given in Table 3, and plotted against atomic number in Figure 31.

From one element to the next, across the row, a unit of charge ($+e$) is added to the nucleus and an electron with charge ($-e$) is added to the valence shell. But only about 35 per cent of the electron density of that electron lies between another given electron (in the same quantum shell) and the nucleus, thus shielding about $+0.35e$ from the added electron (Appendixes 2 and 5).

> What then is the increment in Z_{eff} (for valence electrons) from one element to the next along the row?

Z_{eff} increases by $(1-0.35)$, i.e. by 0.65, in units of e, from one element to the next across the row (as in Fig. 31).

The mutual screening of electrons in the same shell is relatively poor, since they are all at about the same distance from the nucleus.

Thus the effective atomic number increases markedly across the row of the Periodic Table, because the electron shielding (which expresses also the mutual repulsion of the electrons) does not keep pace with the increase in nuclear charge.

37

Here we have the key to understanding our four trends across the row, since the force holding the valence electrons increases with Z_{eff} (see Appendix 3).

As Z_{eff} increases, all the electrons are increasingly pulled in (as we see in the TV programme). Thus, although the atom gets heavier it gets smaller, and its ionization energy, electron affinity and electronegativity all increase (overall) across the row.

This is why we progress from metals on the left, to non-metals on the right, in the Periodic Table.

We can understand the irregularities in periodicities, in terms of Z_{eff}. Let us look at the fourth row of the Periodic Table, in which the transition elements appear between Groups II and III.

What is the increase in Z_{eff} (Table 3 and Figure 31) between calcium and gallium?

This large increase is a result of the increase in Z_{eff} across the transition series itself, as d electrons are added to the atoms.

There is an enormous increase in Z_{eff}, from 2.85 for calcium to 5.00 for gallium.

Predict the effect of the large increase in Z_{eff} on the variation of the electronegativity (use ϵ_p) and of the covalent radius across the fourth row.

ϵ_P increases when Z_{eff} increases across the row, so should increase greatly from Ca to Ga. r_c decreases as Z_{eff} increases (across the row) so should decrease greatly from Ca to Ga.

Figure 25 shows the very large increase in ϵ_P between calcium and gallium (and similarly between strontium and indium in the next row).

Figure 28 shows the very large decrease in r_c from calcium to gallium (and similarly between strontium and indium in the next row).

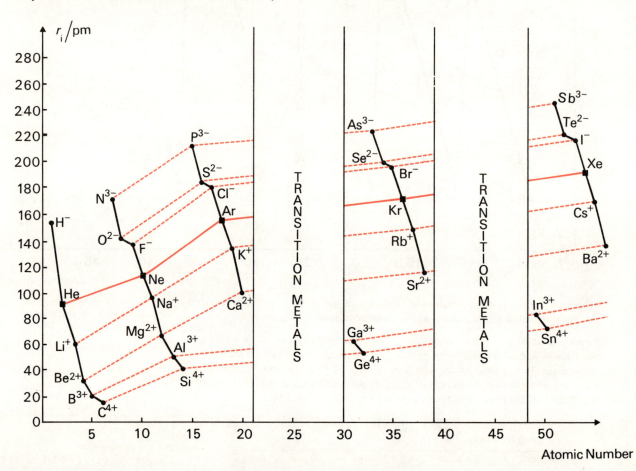

Figure 32 Ionic radii (r_i) of the typical elements. Ions such as B^{3+}, C^{4+}, and Si^{4+}, with high charges and small radii, are included for the sake of completeness, but should not be taken too seriously. Such a high charge density tends to be unstable, and the bonds that are assumed to be ionic (in the determination of these radii) have considerable covalent character. The points ■ show the van der Waals radii of the noble gases, for comparison. We can consider these atoms as ions with zero charge.

Notice also in Figure 32 the large decrease in r_i from Ca^{2+} to Ga^{3+} (and also from Sr^{2+} to In^{3+} in the next row), and in Figure 23 the levelling-off of the ionization energy from calcium to gallium (and from strontium to indium), when by analogy with the previous rows, we should have expected a sharp decrease.

As you can imagine, the chemical properties of gallium and indium, and the elements that immediately follow them, are greatly affected by their position in the Periodic Table—immediately following the transition elements. Because of this well-known 'anomaly', they are called '*post-transition elements*'.

post-transition elements

8.6.6 The periodic variation of ionic radii

We have not yet discussed the changes in ionic radii across the row (Fig. 32), for two reasons. The first is that these are complicated by changes in oxidation number, particularly when this changes sign. The second is that the concept of the effective nuclear charge is helpful in understanding ionic size.

In the TV programme, we show some of the remarkable changes in size that accompany ionization. Thus, compare the covalent radius for lithium (135 pm) with the ionic radius for Li^+ (68 pm). In this instance, the atom loses the 2s shell on ionization; but even when no shell is lost, the change in size is still considerable, as you will see if you compare F and F^-. For lithium the (covalent) radius is halved on ionization; for fluorine it is doubled.

ionic radius

We can understand the periodicity of the ionic radii by considering the series that are connected by black lines in Figure 32, e.g. H^- to C^{4+}, N^{3-} to Si^{4+}, and so on.

> SAQ 13 (Objectives 9, 10)   What have the ions in a given series in common? How does the effective atomic number vary within the series? Compare this with the variation in Z_{eff} across the row for covalently bound atoms. Does this explain the difference we observe in the way r_i decreases within the series, compared with the decrease in r_c across the row of the Periodic Table?

Check your answer with that on p. 54 before you go on.

8.7 Changes down Periodic Groups

A Group is a family of elements which have the same valence configuration of electrons, but which differ in what lies beneath. The elements share the properties that depend on this configuration (e.g. s^1 for the alkali metals), but these properties are modified as we go down the Group, as the number of inner shells increases. But the Group trend that results from this is often irregular, because of irregularities in the way the atoms and the Periodic Table are built up. The greatest 'irregularity' is of course the appearance of the transition elements in the fourth and subsequent rows (and the appearance of the lanthanides and actinides).

We mentioned in Section 8.5 the tendency, as shown in Figure 22, for metallic character to increase down the Groups.

Group trends

> As to the chemical progressions down the Groups, can you pick out some examples from Unit 6?

The increase down the Groups of the oxidation numbers of the highest fluorides, oxides and chlorides was detailed in Unit 6, Section 6.1.3. The increase in the thermal stability of salts of Group I and II metals, relative to a solid product that contains smaller anions, was described in Unit 6, Section 6.3.

8.7.1 Changes in 'atomic' properties down the Groups

The examples just given are related to the increase in atomic and ionic size down the Groups as electron shells are added to the atom, which we illustrate in the TV programme.

These increases are shown by Figures 27, 28 and 32, and documented in your *Data Book*. Figures 23, 25 and 26 show that (irregularities apart) ionization energies and electronegativities decrease down the Group. This, like the increase in size, suggests that the pull of the nucleus on the valence electrons decreases down the Group. But the effective atomic number *increases* down the Group (except for Groups I and II, in which it remains constant after the third row).

How can we reconcile these differences?

As you have read in Appendix 3 (White), the force holding the outer electrons increases with Z_{eff}, but it also *decreases* sharply as the distance of the electrons from the nucleus increases (as we see in the TV programme). Figure 2 (on p. 9) shows that this distance increases as extra shells are added to the atom, and this is the dominant factor here.

Thus the elements tend to progress, down the Groups, from non-metal to metal as the atoms increase in size.

So much for the overall trends. Now let us look at the irregularities. The graphs of ionization, electronegativity, and covalent and ionic radii show that a marked trend is set (down the Groups) for the pre-transition elements in Groups I and II, but the trend is diminished and sometimes reversed for the post-transition elements from Group III onwards (Section 8.6.5).

How is this explained in terms of the Z_{eff} of the valence electrons?

We discussed the consequences of this in Section 8.6.5.

The ionization energy decreases only a little down the Groups, after the large decrease from the second-row to the third-row elements (Figure 23).

Since Z_{eff} increases across the transition series itself, Z_{eff} is greatly increased for the post-transition elements, relative to the first two elements of the row.

The Pauling electronegativity apparently increases in the lower reaches of Group III, and stays constant from silicon onwards in Group IV, although, by all criteria, one would say that lead is more metallic than silicon! The Allred–Rochow electronegativities are more convincing here, since they decrease down Groups III and IV after the third-row elements; but like the Pauling values they *increase* from aluminium to gallium, and from silicon to germanium, for reasons connected with the filling of the d shells (see Section 8.6.5).

A significant 'irregularity' in the Periodic Table, evident in Figures 27 and 28, is that the atoms after lithium in the second row are particularly small. This seems to be due to the small number of inner-shell electrons, only two. We shall see in Unit 9 how important this is in the chemistry of the second row elements.

Summary of trends in the Periodic Table

Summarize briefly the main line of our argument in Sections 8.5 to 8.7, and compare your account with the one below.

Across the row

Across the row in the Periodic Table, the elements progress from metals to non-metals. At the same time there is an overall increase* in the ionization energy, the electron affinity and the electronegativity of the atoms, and a decrease in covalent radius.

We found that all of these progressions could be understood in terms of the increase across the row in the effective atomic number Z_{eff} experienced by the valence electrons. (The effective nuclear charge is Z_{eff}, in units of e.) The decrease in size of the isoelectronic ions across the row, and also the changes in size on ionization, can be understood in terms of Z_{eff}. r_i (for isoelectronic ions) decreases more sharply than r_c across the row because the ions increase in oxidation number, whereas r_c is measured for single bonds.

Z_{eff} increases across the row because shielding of valence electrons from the nucleus by electrons in the same shell is only about 35 per cent efficient. (The shielding is defined to allow for electron repulsion.)

The changes across the fourth and subsequent rows of the Periodic Table are interrupted by the appearance of the transition elements between Groups II and III, i.e. between calcium and gallium. Since gallium, compared with calcium, has

* Irregularities in these properties are related to differences in electron shielding and repulsion in the atoms compared.

a filled d shell as well as an extra valence electron, there is a large increase in Z_{eff} from calcium to gallium. Thus the post-transition atoms and ions are greatly contracted, and have a much higher electronegativity (relative to the pre-transition elements) than we would expect from comparison with the typical elements in the earlier rows.

Down the Group

Down the middle Groups, the elements progress from non-metals to metals, while in the earlier and later Groups the corresponding tendencies are much less marked. The changes down the Groups are rather irregular, due to irregularities in the building up of the electron shells of the atoms. Overall, the tendencies are for the ionization energy and the electronegativity of the atom to decrease (except for the transition and post-transition elements); similarly the covalent and ionic radii increase (this tendency being reduced for the post-transition elements). The changes from the second row to the third tend to be greater than changes between subsequent rows (the second row elements having very few inner electrons).

Thus the atomic and ionic sizes increase down the Groups as more electron shells are added to the atom; the increase in Z_{eff} does not overcome the repulsion of the filled electron shells. Similarly, the ionization energy and electronegativity tend to decrease down the Group with distance of the valence electrons from the nucleus, and increased shielding (repulsion) by inner-shell electrons.

8.8   Hydrogen

Hydrogen seems to be the raw material of the universe. From the spectroscopic evidence, more than 90 per cent of all atoms are hydrogen, and in stars such as the sun, protons (H^+) come together at very high temperatures to build up a-particles (He^{2+}) and heavier nuclei. All the other elements are synthesized in these thermonuclear reactions*, and the energy they produce keeps the stars hot.

cosmic hydrogen

Planets vary in composition. Much of Jupiter is solid, liquid or gaseous hydrogen, but hydrogen accounts for only 15 per cent of the atoms of the Earth's crust, including the hydrosphere. There is some elementary hydrogen in coal and natural gas but, being so light, it escapes from the Earth's gravity. Hydrogen in the form of water gas ($H_2 + CO$) is made in the reduction of steam by red-hot coke. Hydrogen is made also from steam and hydrocarbons, and as a byproduct of the cracking of petroleum hydrocarbons (to improve their octane number), and of electrolyses such as that of brine to make caustic soda.

Hydrogen is consumed in great quantities in the Haber ammonia process, and in other catalytic hydrogenations, such as the hardening of vegetable oils to make fats such as margarine, and the manufacture of methanol from carbon monoxide.

Because of its simplicity, hydrogen is the most versatile of elements. Covalent in most of its compounds, it can also be acidic (solvated H^+), or hydridic (H^-) (Fig. 33). It can form protonic bridges known as *hydrogen bonds*, and also *hydride bridges*, which we discuss in Unit 9.

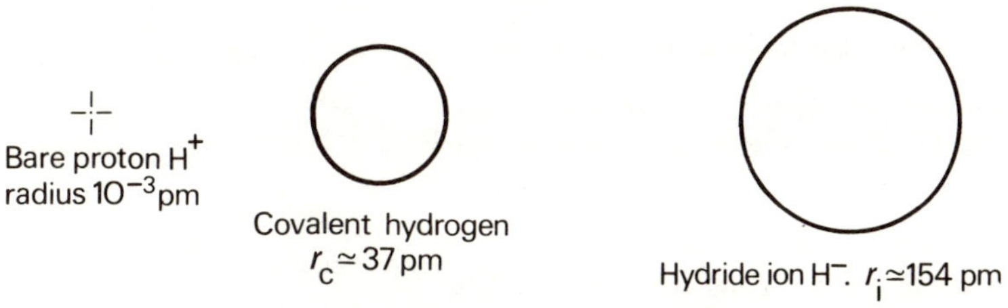

Bare proton H^+
radius 10^{-3} pm

Covalent hydrogen
$r_c \simeq 37$ pm

Hydride ion H^-. $r_i \simeq 154$ pm

Figure 33   The different sizes of hydrogen.

* Prout's hypothesis is vindicated! In the early days of the determination of atomic weights, it became evident that many or most were whole multiples of that of hydrogen. In 1815–16 a London physician, Dr. Prout, formulated the hypothesis that hydrogen is the primordial substance (that the ancient Greek philosophers spoke of) of which all elements are composed. His hypothesis was discredited when more accurate work showed that many atomic weights are not whole numbers (based on H = 1), that of Cl (=35.5) being a serious stumbling block. The discovery of isotopes and fundamental particles vindicated the hypothesis in a metaphorical sense. Now nuclear physics and astronomy have demonstrated the process of synthesis of other elements from hydrogen.

Hydrogen has a middling electronegativity, 2.1 on the Pauling scale. It can oxidize the more electropositive elements, and reduce the more electronegative ones, as shown by the series in Unit 5, Table 1. In the TV programme we show the remarkable differences in size between covalent and ionic hydrogen (Fig. 33).

The bare proton is so small (10^{-3} pm) compared with normal atomic dimensions that the high charge density distorts the electron cloud of other atoms. In aqueous acids the proton is solvated by water, and surrounded by a cluster of hydrogen-bonded molecules, so that the charge is spread. Aquation of the proton is very exothermic, to the tune of 1 120 kJ mole^{-1}.

8.8.1 Hydrides and the Periodic Table

Figure 34 shows the types of hydrides formed by the Periodic Groups. It illustrates neatly the increase in electronegativity across the row.

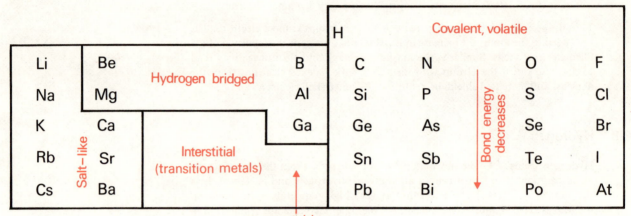

Figure 34   Hydrides and the Periodic Table.

Ionic hydrides M^+H^- and $M^{2+}2H^-$ are formed only by the most electropositive metals, when the metal is heated in hydrogen. They are truly ionic solids: transparent, brittle, high melting and non-conducting. The alkali metal hydrides have the sodium chloride structure (Unit 2).

ionic hydrides

The molten solids show ionic conductance, to give the metal at the cathode and hydrogen at the anode, obeying Faraday's laws.

The ionic hydrides are good reducing agents, and are used to prepare hydrides or reduce various groups in inorganic and organic chemistry. They are very sensitive to water:

$$H^- + H_2O \rightarrow OH^- + H_2(g)$$

Moving across the Table, we come to the *covalent hydrides*, first the hydrogen-bridged ones, which are hydridic ($H^{\delta-}$); then the non-polar ones, such as CH_4; then the protonic ones ($H^{\delta+}$) such as H_2O and HF, as the ligand becomes more electronegative. The hydride bridges in beryllium and boron hydrides will be described in Unit 9.

covalent hydrides

Boron hydrides, now called boranes, have an extensive chemistry (Unit 9).

As we have seen, carbon has a similar electronegativity to hydrogen; the hydrocarbons (such as CH_4) are non-polar, and do not react with polar reagents. In addition, the bonds are strong, and need a sizable activation energy for radical reactions.

After carbon, the hydrides gradually become acidic. Thus the alkali metals displace hydrogen from ammonia to form an amide (see Unit 9):

$$2Na + 2NH_3 \rightarrow 2NaNH_2 + H_2$$

$$\downarrow H_2O$$

$$2NaOH + 2NH_3$$

We compare the Group V hydrides, and make some of them, in the TV programme of Unit 10.

42

Water can act as an oxidizing or a reducing agent, and as an acid or base:

$$2Li + 2H_2O = 2Li^+ + 2OH^- + H_2$$
$$2F_2 + 2H_2O = 4HF + O_2$$

Pure liquid hydrogen fluoride has a very low conductivity, but is a strong acid. In dilute aqueous solution it is a weak acid. This unusual behaviour is explained in Unit 11.

> SAQ 14 (Objectives 8, 11) Describe how the types of hydride formed by the elements of the second row of the Periodic Table illustrate the increase in electronegativity across the row.

Hydrogen bonding

Hydrogen bonding has been described in Units 6 and 7 and in S100, Unit 10. It is a rather weak interaction for, as Table 4 shows, hydrogen bonds are less than one-twentieth as strong as covalent bonds. Only the most electronegative atoms (F, O, N mainly) form hydrogen bonds, and Table 4 shows that the strengths of hydrogen bonds follow the electronegativities of the bonded atoms. This is understandable if we imagine that the electronegative atom, say oxygen in water, leaves a partial positive charge on the hydrogen bonded to it, and this is then attracted to a lone pair on another oxygen atom, since hydrogen bonds (e.g $O-H\cdots O$, $N-H\cdots N$, $O-H\cdots N$ etc.) are usually linear.

hydrogen bonds

Table 4 The strengths of hydrogen bonds

Hydrogen bond	$-\Delta H$ in kJ per mole of hydrogen bond
FH—F in liquid HF	28.4 (cf. 566 for the H—F bond)
OH—O in water	20.9 (cf. 467 for the O—H bond)
NH—N in liquid NH_3	18.4 (cf. 391 for the N—H bond)
CH—O in CH_3CHO	10.9

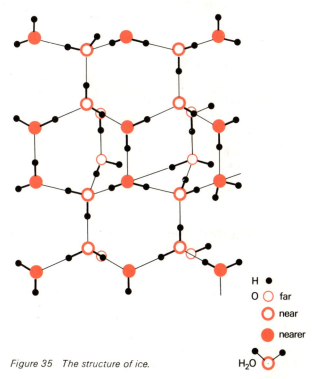

H ●
O ○ far
O near
O nearer
H_2O

Figure 35 The structure of ice.

The structures of water and ice have been intensively studied. In ice (Figure 35) oxygen is tetrahedrally coordinated by hydrogen atoms, two covalently bonded, and two hydrogen-bonded via the lone pairs of that oxygen. The hydrogen lies on the line of centres between two oxygens, linking them loosely together; but it is much closer to the oxygen to which it is covalently bonded. The O—H distance is 100 pm, and the $O\cdots H$ distance 180 pm. In Figure 35 the O—H and $O\cdots H$ bonds are drawn in fixed positions, but they readily switch.

Figure 35 shows that each oxygen is tetrahedrally coordinated by oxygen (with hydrogen in between) and that the oxygen lattice has a hexagonal symmetry. This is well shown in the cover design of S100, Unit 5, which gives the ice structure in projection. This hexagonal symmetry is reflected in snowflake crystals (see Figure 36).

> SAQ 15 (Course Objectives) Explain the tetrahedral coordination of oxygen by oxygen in ice.

The directed bonds give ice its very open structure and low density. Thus, ice floats on water, and the fish in the ponds can live through the winter.

In water, however, there is only short-range structure. The hydrogen bonds break and re-form, with the thermal motion of the molecules. In this system, the proton is apparently extraordinarily mobile. In fact, a proton may join one end of a long hydrogen-bonded chain of molecules, while a proton is lost from the other end (in electrolysis, for example, or a neutralization reaction).

Figure 36   A snowflake.

On all these grounds (the energy relationships included), the hydrogen bond is thought to be mainly an electrostatic interaction, with the proton holding the two electronegative atoms together (the one to which it is covalently bonded, and the other to which it is weakly (hydrogen-) bonded).

In Unit 9, Section 9.9.1, we examine the important part played by hydrogen-bonding in determining the physical and chemical properties of ammonia, water, and hydrogen fluoride.

> SAQ 16 (Objective 12 and Course Objectives) Why is hydrogen-bonding restricted to hydrides of elements in one corner of the Periodic Table?

Interstitial hydrides, which are compounds of the transition metals, form an interesting section of hydrogen chemistry.

interstitial hydrides

As their name suggests, the hydrogen atoms or ions (the charge varies in different hydrides) occupy tetrahedral holes in the metal lattice or sometimes octahedral holes, with lighter metals.* The hydrides are usually non-stoichiometric, more hydrogen being taken up at lower temperatures and higher pressures. They do however have some of the properties of chemical compounds: there is often a considerable heat of formation, structural changes occur at certain compositions, and the magnetic and electrical properties of the metal are often altered. These hydrides are sometimes compared to alloys, with the interstitial hydrogen showing some 'metallic' character.

The hydrogen is often very reactive; thus nickel, platinum and palladium are important catalysts for hydrogenation. Hydrogen diffuses through many metals; it can be purified by passage through palladium, for example.

As background to this hydrogen chemistry, read Chapter 2 in R. C. Johnson's book** (which we shall refer to as Johnson), omitting Section 4.

8.8.2   Bond energies in the hydrides

The strengths of the bonds that hydrogen forms with the other elements show an interesting periodicity (Fig. 37).

bonds to hydrogen

There is an overall increase in bond energy across the row as Z_{eff} of the ligand increases, with 'extra' stability for the ionic bonds M^+H^- and H^+A^-. The minima at Groups II and V seem to reflect the additional stability of the s^2 and s^2p^3 configurations in the free ligand atoms; since the bond energy term (*Data Book*, Section 8) is defined relative to the free atoms.

Down the Groups, the decrease in the bond energy term probably reflects the increasingly poor overlap of the small 1s orbital of hydrogen with the diffuse

---

* A tetrahedral hole is the space in the middle when four close-packed balls form a tetrahedron, just as an octahedral hole is the space in the middle when six close-packed balls form an octahedron. Both are present in the close-packing of spheres (Unit 3) as you can verify with the Home Experiment Kit.

** R. C. Johnson (1966) *Introductory Descriptive Chemistry*, Benjamin.

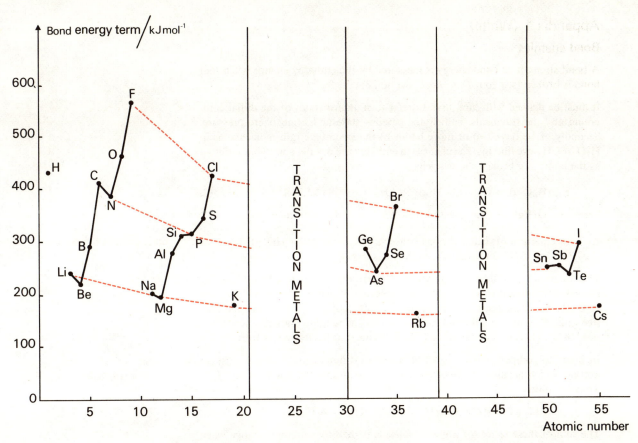

Figure 37 *The strengths of bonds with hydrogen*

orbitals of the larger atoms, and the increasing disparity in the energies of the overlapping orbitals (Section 8.4.1). Thus hydrogen forms strong bonds with smaller atoms (as in the H_2 molecule), and weaker bonds with larger atoms.

SAQ 17 (Objective 11, and Course Objectives) List brief arguments for and against putting hydrogen in

(a)   Group I
(b)   Group IV
(c)   Group VII

Appendix 1 (White)

Bond energies

A bond strength or bond energy is measured by the enthalpy change when that bond is broken (see your *Data Book*, Section 8).

It must be defined with care, first, to refer to standard states of the initial compound and the fragments, namely the gaseous state at 1 atmosphere pressure. Secondly, if we have two or more bonds of the same type in a molecule, as in H_2O or CH_4, we find that the changes in enthalpy are usually somewhat different as the bonds are broken successively.

$$\text{e.g. } H_2O(g) \rightarrow OH(g) + H(g) \qquad \Delta H_m^\ominus = +494 \text{ kJ mol}^{-1} \qquad (2)$$

$$OH(g) \rightarrow O(g) + H(g) \qquad \Delta H_m^\ominus = +431 \text{ kJ mol}^{-1} \qquad (3)$$

ΔH_m for equation 2 is called the *bond dissociation energy* of the HO—H bond, or $D_m(HO\text{—}H)$. $\Delta H_m^\ominus$ for equation 3 is thus $D_m(O\text{—}H)$.

bond dissociation energy

The reason for this difference is that the OH group is different when it is a free radical, and when it is part of the H_2O molecule. When the first H is lost from H_2O, the electrons rearrange to the appropriate configuration for the diatomic free radical. Sometimes the radical changes its shape quite markedly when it is free; thus CH_3 is pyramidal in the CH_4 molecule, but flat when it is free.

free radical

In order to compare the strengths of bonds in different molecules, we use an average which is called the *bond energy term*, or simply the bond energy, E_{AB}. Thus for water

bond energy term

$$E_{m, OH} = \tfrac{1}{2}[D_m(HO\text{—}H) + D_m(O\text{—}H)] = 463 \text{ kJ mol}^{-1}$$

The sum of these terms for a given molecule is then the enthalpy change when the molecule is atomized (with molecule and atoms in their standard states).

SAQ 18 (Objective 1)   What is the value of E_{NH} given that for

$$NH_3(g) \rightarrow NH_2(g) + H(g), \qquad \Delta H^\ominus = D(NH_2\text{—}H) = 448 \text{ kJ mol}^{-1}$$
$$\text{and similarly } D(NH\text{—}H) = 368 \text{ kJ mol}^{-1}$$
$$D(N\text{—}H) = 356 \text{ kJ mol}^{-1}?$$

All quantities are molar.

When is the bond energy term necessarily equal to the bond dissociation energy?

If we take the enthalpy of atomization of *any* molecule as equal to the sum of the bond energy terms, we are assuming additivity of *different* bond energy terms, i.e. that E_{AB} is constant in different molecules containing this bond. Thus for $CH_3OH(g) = C(g) + 4H(g) + O(g)$.

$$\Delta H_m^\ominus = 3E_{CH} + E_{CO} + E_{OH}$$

where E_{CH} is obtained from CH_4 and E_{OH} from water. (Relationships of this type are used to obtain E_{CO}, in practice.)

This additivity is, of course, subject to the limitations that are becoming familiar to us from our discussion of bond lengths, for example. It applies only to fixed valencies of the atoms, specified bond multiplicities (single, double, etc., bonds), and so on. It is particularly useful in organic thermochemistry, because of the large range of similar compounds. In our discussions, we shall frequently refer to bond energy terms, or to bond dissociation energies for diatomic molecules.

46

Appendix 2 (White)

The effective nuclear charge

As we have seen in Section 8.2, much progress has been made in theoretical chemistry by the extension to more complicated atoms of information gained from a study of the hydrogen atom.

The Schrödinger equation can be solved exactly for the hydrogen atom to give, for example, the energy of the electron, equal to the ionization energy I_H that we measure for the 1s electron. The same value for I_H can be obtained if we apply classical (Newtonian) mechanics (S100, Unit 4) to the Coulomb attraction of the two oppositely charged bodies, the nucleus and the electron, but assume that these are a fixed distance (r) apart. (There is more about this in Appendix 5 (Black).) If then we use the measured value of I_H to find r, this turns out to be the same as the 'distance of maximum probability of the electron being found' that is given by wave mechanics (cf. Section 8.1 and Figure 2).

ionization energy from the Coulomb law

You may well ask 'is the electron a wave or is it a particle? How can a wave obey the Coulomb law? How can a particle be on both sides of the nucleus at once?' The answer is simply that a model is justified so long as it is useful. As Bragg said, 'We teach that electrons are waves on Mondays, Wednesdays and Fridays, and that electrons are particles on Tuesdays, Thursdays and Saturdays'. (He was describing courses with three lectures a week.)

More complex atoms than hydrogen can be viewed in this way, but whereas the hydrogen atom is a two-body problem, an N-electron atom is an (N + 1)-body problem, in which the unlike charges (nucleus and electron) attract, and the like charges (electrons) repel, according to the Coulomb law. But there is no exact (i.e. algebraic) solution to three-or-more-body problems, whether the bodies are subatomic, cosmic, or any other size, and whatever the nature of the forces between them.

So we look for good methods of approximation.

We mentioned that in the carbon atom, with atomic number (Z) equal to 6, the number of pair-wise interactions, given by $\frac{1}{2}Z(Z + 1)$, is equal to 21. If we replace the nuclear charge, which is equal to $+ Z$ (in units of e, the electronic charge), by Z_{eff} (the effective nuclear charge for a given type of electron) which takes account of screening as explained below, then the number of pair-wise interactions is reduced to 6, the number of electrons each interacting with the nucleus. This also takes care of the mutual repulsion of the electrons, and makes it easier for us to visualize the different effects of the different types of electron (1s, 2s, etc.).

effective nuclear charge

We can explain the effective nuclear charge as follows: If we remove an electron from an atom of atomic number Z, to leave a cation with charge $+1$, this charge as experienced by the external electron is equal to the nuclear charge $+Z$ less the charge ($Z - 1$) of the electrons shielding it (since at external points, the charge on a spherical shell acts as if concentrated at the centre).* We can describe this as 100 per cent screening, in the sense that each electron screens one positive charge on the nucleus.

But if that electron returns to the valence shell of the atom, it experiences an effective nuclear charge Z_{eff} that is greater than $+1$, as you can see from Table 3 and Figure 31. There are two main reasons for this:

1   The valence shell has a certain thickness, and some of the charge of the other valence electrons does not lie between this electron and the nucleus. (A spherical shell of charge exerts no resultant field at internal points.*) It is estimated (Appendix 5) that the screening of an electron in the valence shell by another in the same shell is only 35 per cent effective. This means that, from one element to the next in the row of the typical elements, there is an extra charge of $+1$ in the nucleus, but the extra electron only screens 0.35 of that charge from another electron in the same shell.

shielding by electrons in same shell

Figure 31 shows Z_{eff} increasing in steps of 0.65 across the row of the typical elements.

---

* This is a theorem of electrostatics; the spherical shell is uniformly charged.

2 The valence electrons penetrate the inner shells, as we can see by comparing Figure 2 with Figure 5. The small inner lobe of the 3s orbital, for example, is within the $n = 1$ shell.

> What magnitude of effective nuclear charge does a 1s electron experience?

The small amount of screening is about $0.3e$ in fact: a 1s electron in the sodium atom 'sees' a nuclear charge of $10.7e$. Thus the small part of the 3s electron distribution that is in the inner lobe 'sees' a very large Z_{eff} and this raises the average value for the orbital as a whole.

The Z_{eff} approximation enables us to treat a many-electron atom as a set of 'two-body problems'. One 'body' is a central core consisting of the nucleus and $(N - 1)$ electrons, and the other is the electron we are interested in. This is a valence electron, since our present concern is with structure and bonding. The size of an atom or ion (r_w, r_c or r_i) is determined by the radius of the most loosely-held electron; the electronegativity of the element depends on the force with which that electron is held by the core.

> Which Group of the Periodic Table is best suited to this description of the atom as core plus valence electron?

For the alkali metals, the nucleus and inner shells form the core.

> SAQ 19 (Objective 10)   Find from Table 3 the values of Z_{eff} for lithium and sodium valence electrons. Why is each greater than one?

The force with which the valence electron is held, in a hydrogen-like atom, is given approximately by the Coulomb law (S100 Unit 4, Section 4.2.2) as

$$F = -\frac{k\,Z_{eff}\,e^2}{r^2} \qquad (4)$$

where k is a constant.

For the value of r in this equation we can use r_c, as for the Allred-Rochow electronegativity (Section 8.6.3); or we can use the radius of maximum electron density as given by (approximate) solutions of the wave equation (Section 8.1).

Equation 4 shows that the force increases as Z_{eff} increases; but it decreases more sharply with increase in r, since the r term is squared.

Across the row, Z_{eff} increases steeply, the increasing force pulls the electrons in, and r_c decreases. The repulsion of electrons in the same shell is outweighed by the increased nuclear charge.

Down the Group, r increases as extra electron shells are added, and this tends to decrease F, in spite of the increase in Z_{eff} in most Groups. The repulsion of the filled quantum shells outweighs the increased attraction due to increase in the nuclear charge down the Group.

Appendix 5 (Black) describes how Z_{eff} values are obtained from ionization energies, and by Slater's rules.

Appendix 3 (Black)

Representations of angular functions

Many textbooks represent the angular functions using *polar diagrams*. These are best explained with a few examples. Figure 38 shows s and p polar diagrams.

polar diagrams

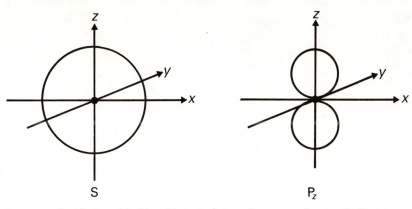

S   P$_z$

Figure 38   Polar diagrams.

For example, the 1s orbital is spherical; ψ_a has the same value in all directions. In a polar diagram, the value of ψ_a in any direction is defined by the length of the line from the nucleus to the surface of the polar diagram in that direction. For the s orbital this is a sphere because the lines defined are radii and these have a constant value irrespective of direction.

For p orbitals, the polar diagram consists of two spheres in contact at the nucleus. Clearly the value of ψ_a now depends on the direction. ψ_a is zero in all directions in the xy plane; this plane is a nodal surface. ψ_a has a maximum value in the z direction. The other two p orbitals have nodes in the yz and xz planes, with lobes directed along the x and y axes.

You should not confuse these polar diagrams with the boundary surfaces that we use in this Unit. Polar diagrams represent the angular functions, not the atomic orbitals.

Appendix 4 (Black)

The relative stabilities of different electronic configurations in atoms

Irregularities in the plot of ionization energy or electron affinity against atomic number show that the electronic configurations s^2 and s^{2}p^3 are slightly more stable than we would expect if these quantities increased steadily across the row. You are familiar with the special stability of the filled shell s^{2}p^6. It is interesting that the s^2 configuration is that of a filled subshell, with $l = 0$, and that the p^3 configuration is that of a symmetrically half-filled subshell with $l = 1$; since by Hund's rule (Section 8.2), the three p electrons occupy the three different p orbitals. The symmetrical electron distribution reduces electron repulsion.

If we think in terms of the local minima, we find that the s^{2}p and s^{2}p^4 configurations are slightly less stable than we would expect (if the ionization energy and electron affinity increased steadily across the row). The lower stability of the s^{2}p than the s^2 configuration reflects the looser binding of the p electron than the s electron (in the same shell). The clue to this can be seen in Figure 2*. This shows that the s electron, compared with the p electron, has a greater probability of getting close to the nucleus, as in the inner lobe of the radial distribution. The s electron therefore experiences more of the nuclear charge than the p electron does, and is more tightly held. In addition, it shields the p electrons rather better than they shield each other.

In the s^{2}p^4 configuration, the fourth p electron is somewhat repelled by the p electron with which it has to share an orbital, and so it is that much easier to remove; and similarly for the fifth and sixth p electrons in the s^{2}p^5 and s^{2}p^6 configurations.

But, as we have seen in Section 8.4.2, the distinction between s and p electrons is usually less marked for atoms in molecules, than for free atoms.

* Figure 2, strictly speaking, refers to the hydrogen atom, for which s and p orbitals with the same number of n are degenerate. But in all other atoms the ns level is below the np level, for the reason given.

Appendix 5 (Black)

Estimation of the effective nuclear charge; Slater's rules

At the beginning of Appendix 3 we mentioned that the ionization energy of the hydrogen atom (I_H) can be calculated exactly by classical mechanics (assuming a fixed radius)* and by wave mechanics. Both give

calculation of ionization energy

$$I_H = \frac{k Z e^2}{2r}$$

where k is a constant, and $Z = 1$ for hydrogen.

r_H, the distance of maximum probability of finding the 1s electron, is given by the Schrödinger equation as 53 pm, and this value is obtained by physical measurements of the hydrogen atom radius, e.g. in diffusion experiments. As we mentioned before, this is also the value that fits the classical equation.

The Schrödinger equation gives in addition (cf. S100 Unit 30, Section 30.4.2 and Appendix 6):

$$I_H = \frac{2 k^2 \pi^2 Z^2 e^4 m_e}{n^2 h^2}$$

$$= \frac{k' Z^2}{n^2} \qquad (5)$$

where k' assimilates all the constants.

n is the principal quantum number of the electron
h is Planck's constant.
m_e is the mass of the electron

The hydrogen atom in its ground state has $n = 1$, so the constant k' is equal to I_H.

For a hydrogen-like atom such as lithium we can write an analogous equation to equation 5 to give the first ionization energy as:

$$I_{Li} = \frac{k' Z_{eff}^2}{n^2} = \frac{I_H Z_{eff}^2}{n^2} \qquad (6)$$

In this way, we can use measured values of I_H and I_{Li} to determine an effective atomic number for the valence electron in the lithium atom, and hence discover how well the two inner electrons screen the nuclear charge from the valence electron, since

$$Z_{eff}\, e = Ze - \text{screening by inner electrons}$$

> SAQ 20 (Black) (Objective 10)   Calculate a value of Z_{eff} for the lithium valence electron, using equation 6 and ionization energy values from your *Data Book*, Section 9.1.
>
> What is the charge screened by the 1s electrons?

The values of Z_{eff} in Table 3 and Figure 31 are given by Slater's rules, which are described below.

Slater's rules for the effective atomic number

J. C. Slater, an American physicist who has made important contributions to theoretical chemistry, used the effective atomic number to extend the mathematical description of the hydrogen atom to polyelectronic atoms or ions. He found that only the alkali metals are sufficiently hydrogen-like to give a value of Z_{eff} that expresses other properties of the atom than the ionization energy used to derive it. More generally useful values were derived from Slater orbital functions. These are wave functions developed by Slater, which reproduce well the observed *energies* of electrons in atoms, and also give orbital sizes that match observed atomic and ionic radii.

Slater used Z_{eff} and an effective principal quantum number n as parameters that could be varied to give the best fit with the self-consistent field (Section 8.2)

---

* The angular momentum is quantized (S100, Units 29 and 30).

values of the radial function. Since Slater orbitals are widely used, he has given a set of rules, which we discuss below, for finding Z_{eff}.

Slater Z_{eff} values match the results of the self-consistent field method, in which the charge distribution of the electrons (minus the one under consideration) is spherically averaged. Since s orbitals are spherically symmetric and p orbitals are not (see Figs. 3 and 4), spherical averaging removes this distinction. Table 5 shows that Slater Z_{eff} values are the same for s and p electrons in the same quantum shell.

Because the distinction between s and p electrons in the core is removed, Slater Z_{eff} values vary smoothly across the row of the Periodic Table, like covalent radii, but unlike ionization energies (cf. Appendix 4).

Slater's rules for Z_{eff}

Table 5 Slater's rules

| | | Electron under consideration | | |
		1s	2s, 2p	3s, 3p
Screening effect of the other electrons	Subtract for each 3s, 3p electron			0.35
	Subtract for each 2s, 2p electron		0.35	0.85
	Subtract for each 1s electron	0.30	0.85	1.0

The effective nuclear charge $Z_{eff}e$, as seen by a particular electron, is obtained by subtracting from Ze the portion of the nuclear charge screened from that electron by the other electrons. The rules apply also to ions, even to atoms or ions with inner electrons missing, as in X-ray spectroscopy. No screening is counted for electrons outside the shell of the one considered. The values are unreliable for $n \geqslant 4$.

Thus for valence electrons with $n = 2$, the screening by the 1s electrons is 85 per cent efficient, if we define efficiency as the percentage of unit nuclear charge screened by one electron.

> How efficient is the screening of valence electrons with $n = 3$, by electrons with $n = 2$, and by electrons with $n = 1$? Obtain your answers from Table 5.

We are interested in Z_{eff} for valence electrons, and will work some examples:

For lithium, $Z_{eff} = 3 - (2 \times 0.85) = 1.3$.

This is close to the value for Z_p that we obtained by calculation, in SAQ 19.

For neon, $Z_{eff} = 10 - (7 \times 0.35) - (2 \times 0.85) = 5.85$.

For sodium, $Z_{eff} = 11 - (8 \times 0.85) - (2 \times 1) = 2.2$.

> SAQ 21   Work out Z_{eff} for valence electrons on carbon and chlorine, and check your answers with Table 3. (Remember that the electron does not screen itself.)

Slater's rules express these observations:

1   The screening of the nuclear charge from the valence electrons by electrons in inner shells is roughly constant, for atoms in the same row.

2   s and p electrons in the same shell screen each other approximately equally.

3   The screening of s and p electrons by others in the same shell is not very efficient, i.e. 35 per cent.

> Can you give a simple physical interpretation of these observations in terms of quantum shells?

The screening is 85 and 100 per cent respectively. As mentioned above, the mutual repulsion of the electrons is covered by these terms.

1 arises because most of the first quantum shell is nearer the nucleus than the second shell.

2 and 3 arise because s and p electrons with the same quantum number are at about the same (mean) distance from the nucleus. Figure 2 shows that the s and p shells, for the same n, have similar mean radii.

Answer to exercise (Section 8.1)

1   The atomic spectrum of hydrogen is a line spectrum. This is direct experimental evidence of energy levels in the hydrogen atom; in the flame the electron in the atom is excited or jumps from one energy level to another. We label these energy levels with the quantum number, n (the principal quantum number). The possible values of n are 1, 2, 3, 4, etc.

According to the laws of electrostatics (Coulomb's law), the energy of the electron decreases as it approaches the nucleus. The most stable (lowest energy) state of the electron is given the value $n = 1$ and is called the ground state.

Read S100, Unit 6, Section 6.5 for a further revision of these ideas.

2   The atomic spectra of all atoms are line spectra. The successive ionization energies of atoms also show that the electrons have specific energies.

The values of the successive ionization energies show also that the electrons are grouped into shells. Remember that a low ionization energy corresponds to a relatively large electron-nucleus distance, and a high ionization energy corresponds to a short electron-nucleus distance.

S100, Unit 7, Section 7.1.1 describes how the ionization energies of sodium reveal the presence of three electron shells. Within a particular quantum shell, say $n = 2$, some electrons have different energies. We therefore conclude that electrons occupy sub-shells. The principal quantum number shells, denoted by the quantum number n, are divided into a number of sub-shells denoted by the azimuthal quantum number, l. Energy sub-shells are discussed in S100, Unit 7, Section 7.2.

3   (i) When a beam of electrons is directed at two parallel slits placed in front of a photographic film a pattern of alternating dark and light bands is recorded on the film (Young's interference experiment).
(ii)   As shown by Davisson and Germer a beam of electrons is diffracted by a crystalline solid.
(iii)   The regular interatomic distances within molecules also produce diffraction effects and this is the basis of gas electron diffraction (Unit 2).

4   The electron is confined by the attractive force of the nucleus but is free to move in three dimensions. So we expect that the electron wave in the atom should be described by three quantum numbers. You should read S100, Unit 30, Section 30.5 for a revision of electron waves in atoms.

The energy levels of the electrons within atoms are determined by the values which the three quantum numbers take. In fact when atomic spectra are investigated with very high resolution spectrometers, a fine splitting of some lines is observed. This can only be accounted for by a further splitting of the energy levels, and so a fourth quantum number is necessary to describe electron waves. This is the spin quantum number, s.

s takes the values $+\frac{1}{2}$ and $-\frac{1}{2}$ and the 'spin' of the electron is represented by an arrow, $\uparrow$ or $\downarrow$. However, the physical picture of a spinning electron forms no part of the wave model and in fact a theoretical explanation of the fine structure of the spectra has been proposed. Nevertheless, the term spin conveniently describes the quantization and is retained.

The idea of a spinning electron is based on an experiment performed by Stern and Gerlach in 1922. They found that a beam of silver atoms was split into two beams by a very strong and uneven magnetic field. This surprising result is explained by the magnetic properties which result from the spin of the unpaired electron of the silver atom. We call this behaviour in a magnetic field *paramagnetism* and use the magnetic phenomenon as a test of unpaired electrons.

paramagnetism

SAQ answers and comments

SAQ 1   In the hydrogen atom the energy of the orbital depends on the value of the principal quantum number n. The value of n, which controls the number of possible l values, also controls the degeneracy of the energy levels (i.e. n also controls the number of orbitals with the same energy).

The electron distribution depends on both n and l. The size of the orbital depends on n and its shape on l.

SAQ 2   Using the aufbau procedure, the Pauli principle and Hund's rule, the orbital occupancy is given by the electronic configuration $1s^2 2s^2 2p_x^2 2p_y 2p_z$. The three 2p orbitals are degenerate and Hund's rule says that where possible electrons are unpaired.

SAQ 3   He and Li have spherical electron distributions; the distribution in O is predicted to be directional. In $He(1s^2)$ and $Li(1s^2 2s^1)$, only the spherical orbitals are occupied. The occupied p orbitals in O have directional properties. However, you should remember that we have no evidence that atoms have any shape but spherical.

SAQ 4   An atom containing unpaired electrons is deflected by the magnetic field because of the electron spin. He has no unpaired electrons and is undeflected. Li and O have one and two unpaired electrons respectively and the atomic beam would be split as in the Stern–Gerlach experiment.

SAQ 5
1   A σ bonding orbital.
2   No molecular orbital.
3   No molecular orbital.
4   A π bonding orbital.

Look at Figures 14 and 15 for verification of this.

SAQ 6   The ions H_2^+ and He_2^+ have one and three valence electrons respectively. By the aufbau procedure these go into the lowest energy orbitals, and the occupancy is shown in Figure 39.

Both ions have a net excess of one bonding electron; the bond order is $\frac{1}{2}$ in both ions. The theory therefore predicts these ions are stable.

In fact they have been observed spectroscopically and their bond dissociation energies are a little over half the value for H_2 in which the bond order is one. Bond lengths, not surprisingly, are greater than in H_2 : 106 pm for H_2^+, 108 pm for He_2^+. In general a decrease in bond order (number of net bonding electrons) gives a weaker bond which is correspondingly longer.

These ions illustrate a fundamental difference between molecular orbital and Lewis theories. Despite the tendency of electrons to pair (according to the Pauli principle), the basic 'unit' of molecular orbital theory is the single electron bond.

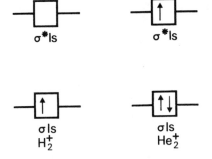

Figure 39   Orbital occupancy in H_2^+ and He_2^+

SAQ 7   The valence shell electron configuration of P is $3s^2 3p_x 3p_y 3p_z$. According to molecular orbital theory, the P atom, like the N atom in NH_3, uses the half filled p_x, p_y and p_z orbitals in linear combination with the 1s orbitals of H. The result is a predicted bond angle of $90°$ (cf. Section 8.4).

In H_2S there are eight valence shell electrons, so there are four repulsion axes to consider. Electron pair repulsion theory thus predicts a bond angle of $109.5°$ (tetrahedral), or slightly less if we allow lone pairs to be more effective than bonding pairs (Unit 7). This adjustment needs to be drastic in the case of PH_3. Obviously molecular orbital theory makes a more accurate prediction for this and analogous molecules, e.g. $AsH_3(91.5°)$ and $SbH_3(91.3°)$.

SAQ 8   These electrons occupy valence orbitals not involved in bond formation – non-bonding orbitals. They are therefore the lone pair electrons analogous to the lone pair in NH_3.

SAQ 9   The linear combination of atomic orbitals is made in a similar way to that for methane (Section 8.4). Four equivalent bonding molecular orbitals are formed from the four 1s orbitals of H and the 3s and three 3p orbitals of Si. The result is a tetrahedral molecule. Alternatively we may say that the Si atom is sp^3 hybridized.

SAQ 10   You should have obtained the first part of Figure 26. (The correspondence to the Pauling electronegativity can be improved if the values are fitted to $\epsilon_{Li} = 1$ as well as to $\epsilon_F = 4$, i.e. by altering the scale.)

SAQ 11   You should have obtained:

Δ_{Si-O} as 280 kJ mol^{-1}, giving $\epsilon_O - \epsilon_{Si} = 1.7$, cf. $3.5 - 1.8 = 1.7$ from Table 3.

Δ_{C-O} as 114 kJ mol^{-1}, giving $\epsilon_O - \epsilon_C = 1.1$, cf. $3.5 - 2.5 = 1.0$ from Table 3.

This example shows how the relatively high bond energy term of the Si—O bond (compared with the C—O bond, say) correlates with the relatively large electronegativity difference between silicon and oxygen. Evidence of the strength of the Si—O bond, relative to the alternative arrangements open to these atoms, is all around us, in silica and silicates.

SAQ 12

$$\epsilon_{AR} = \frac{3590 \times 5.75}{(99)^2} + 0.74 = 2.85$$

This is fairly close to the Pauling electronegativity, which is 3.0 for chlorine. There is closer agreement elsewhere in the Periodic Table, in the second row for example, as you can check for yourself if you wish.

SAQ 13   In each series the ions are isoelectronic, that is, they have the same total electronic configuration. This configuration is that of the noble gas element, you may have said.

In each series, Z_{eff} increases in steps of 1 unit, as compared with 0.65 for the covalently bound atoms across the row (Table 3). In this way we can understand the much greater decrease across a 'row' of isoelectronic ions, compared with the decrease in covalent radius across a row of the Periodic Table. Where there is a sharp decrease in radius, as from Mg^{2+} to Be^{2+}, Al^{3+} to B^{3+}, or Si^{4+} to C^{4+}, this correlates with greatly increased covalent character in the bonds formed by the smaller 'ion'.

SAQ 14   In the hydride formed by lithium, which is the most electropositive element of the second row, hydrogen accepts an electron to form the hydride ion H$^-$, and the compound is salt-like Li$^+$H$^-$. It has the NaCl structure, and reacts with water to give hydrogen gas:

$$H^- + H_2O \rightarrow OH^- + H_2(g)$$

Beryllium and boron form covalent hydrides, but the hydrogen in them is still hydridic, $H^{\delta-}$, giving hydrogen gas with water.

Carbon has a similar electronegativity to hydrogen, and forms covalent hydrides such as the hydrocarbons, CH_4 and C_2H_6, in which the C—H bond is non-polar. (The hydrogen in acetylene is acidic, however, and forms acetylides with reactive metals.)

As the elements become more electronegative across the row, the hydrogen becomes increasingly 'protonic' $H^{\delta+}$, and replacement is possible by an increasing number of metals as we move from NH_3 to H_2O to HF:

$M + NH_3 \rightarrow MNH_2 + \frac{1}{2}H_2(g)$ for Group I metals
$M + H_2O \rightarrow MOH + \frac{1}{2}H_2(g)$ for metals of Groups I and II (Unit 5, Table 1)

SAQ 15   We can explain the tetrahedral coordination of oxygen by oxygen (with hydrogen in between) in the ice lattice, by mutual repulsion of the four electron pairs in the valence shell of oxygen (see Unit 7). The oxygen octet consists of two bonding pairs of electrons forming OH bonds, in which the hydrogen is hydrogen-bonded to oxygen in other molecules, and two lone pairs, each of which forms a hydrogen bond O $\cdots$ H with hydrogen in a neighbouring molecule. Since the hydrogen bond O—H $\cdots$ is linear, this gives tetrahedral coordination of oxygen by oxygen (with hydrogen in between).

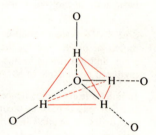

SAQ 16   Hydrogen bonding is restricted to hydrides of the most electronegative elements, in the top right-hand corner of the Periodic Table; that is, of fluorine, oxygen and nitrogen, mainly. These atoms all carry lone pairs and, being electronegative, their covalent bond to hydrogen is a polar one. The hydrogen thus carries a partial positive charge:

and is attracted to a lone pair on oxygen in a neighbouring molecule.

Evidence for this electrostatic description of the hydrogen bond includes the geometry of the arrangement (the linearity of the hydrogen bond (e.g. O—H $\cdots$ O), the tetrahedral coordination of oxygen by hydrogen and oxygen in ice, and so on); and also the weakness of the interaction, about one-twentieth of strength of the corresponding covalent bond.

The most electronegative elements are found in the top right-hand corner of the Periodic Table (if we ignore the noble gases which don't form compounds), where the effective nuclear charge, ionization energy and electron affinity are high, and the atomic and ionic radii low.

SAQ 17   Arguments in favour of hydrogen being added to:

(a)   Group I: like the alkali metals, hydrogen has only one valence electron. Thus singly bonded diatomic molecules H_2 and M_2 are stable in the gas phase; and cations H^+ and M^+ are formed. On the other hand, the Group I elements have four orbitals in the valence shell while hydrogen has only one (since its orbitals with $n > 1$ are too high in energy to be used.) The Group I elements therefore have solid metallic structures at room temperature. They are also very reactive with air, moisture, etc. H_2 is much more strongly bonded than M_2, and is relatively inert (except in free radical reactions such as combustion, which is discussed in Unit 9).

The hydrogen ion H^+ is simply a proton, and much too small to form ionic lattices as M^+ does. Thus HCl is covalently bonded and a gas at room temperature, in contrast to MCl.

(b)   Group IV. Hydrogen and carbon atoms are alike in having a half-filled valence shell. Thus hydrogen and carbon have similar electronegativities, and CH_4 is non-polar (not acidic or basic). Both hydrogen and carbon tend to form covalent bonds. C_2 molecules are stable in the gas phase at higher temperatures.

On the other hand, carbon has four vacancies in the valence shell as against one for hydrogen. Thus carbon cannot form ions as hydrogen does. Carbon is commonly tetravalent, and C_2 is multiple-bonded. The normal form of carbon is a solid with bonds in three dimensions. Hydrogen can never form more than one normal covalent bond (where 'normal' means a two-centre, i.e. two-atom electron pair, bond). In hydrogen bonds and hydride bridges (Unit 9) three atoms are held by the same electron pair.

(c)   Group VIII. Hydrogen resembles the halogens in being one electron short of a filled shell. Thus hydrogen and the halogens form stable diatomic molecules which are singly bonded, as well as single bonds with non-metals such as C, S, P. Hydrogen and the halogens form anions H^- and X^-, LiH and NaH being salt-like, with the NaCl structure.

On the other hand, the hydride ion H^- is a very strong reducing agent, and reacts with water (it is very sensitive to moisture) to give hydrogen, whereas the halide ions are of course stable to water. The halogens heavier than fluorine can have higher valencies than one, as in the chlorate (ClO_3^-) and perchlorate (ClO_4^-) ions.

SAQ 18 $E_{NH} = \frac{1}{3}(448 + 368 + 356) = 391$ kJ mol^{-1}
$E_{AB} = D(A - B)$ only when AB is a diatomic molecule or radical.

SAQ 19 Z_{eff} is 1.30 for the lithium valence electron. It is greater than 1, because part of the 2s electron density lies in the inner lobe of the orbital (Fig. 2), which lies within the 1s orbital, and so experiences the full nuclear charge of $+3e$, less the small amount shielded by the 1s electron.

Thus the average Z_{eff} over the whole orbital is greater than 1. For the sodium 3s electron, Z_{eff} is 2.20 for the valence electron. This is because the 3s orbital penetrates the $n = 2$ and $n = 1$ shells, to the extent shown by Figure 2, and when inside the $n = 1$ shell, the 3s electron experiences almost the full nuclear charge of $11e$.

SAQ 20

$$519 = \frac{1314\, Z_{eff}^2}{2^2} \text{ giving } Z_{eff} = 1.26$$

This is slightly different from the value given by Slater's rules.

$$Z - Z_{eff} = \text{nuclear charge screened by the 1s electrons}$$
$$= (3 - 1.26)e$$
$$= 1.74e$$

Each inner electron screens $0.87e$ nuclear charge.

That the 1s electrons are only 87 per cent efficient in this screening is due to the penetration of the 1s shell by the inner lobe of the 2s orbital (Fig. 2).

SAQ 21   See Table 3 on p. 37.

Unit 9 Elements of the Lithium Row; Orbitals, Part 2

Contents

Objectives

When you have completed this Unit, you should be able to:

1 Define and recognize definitions of, and recognize correct uses of the terms and principles in Table A.
(SAQ 4)

2 Given the relevant sequence of energies of molecular orbitals or an energy level diagram for diatomic molecules and ions of the second period (Li to Ne), determine the following:
(a) whether the molecule is stable relative to the separate atoms;
(b) the bond order of the molecule or ion;
(c) whether the molecule is paramagnetic;
(d) the electronic configuration of the molecule.
(SAQs 1, 2, 3)

3 Rationalize experimental measurements of bond lengths, bond energies and paramagnetic properties using molecular orbital energy diagrams.
(SAQ 3)

4 Recognize instances where hybridization (sp, sp^2 or sp^3) can account for known molecular geometry.
(SAQs 4, 5, 6, 7, 18)

5 Recognize instances where multiple bonding (σ and $p\pi$) is likely to influence molecular stability and geometry.

6 Identify chemical and physical properties that are characteristic of second-row elements and their compounds (as opposed to those of subsequent rows) in the following instances:
(SAQs 9–19, 22–25)
(a) lithium, compared with the other alkali metals; the electrochemical series of the alkali metals;
(SAQs 9, 11, 25)
(b) beryllium, compared with the alkaline earth metals; the bonding in compounds of beryllium;
(SAQs 10, 11, 24)
(c) boron, compared with aluminium and carbon; the boron halides, borides, borates and simple boranes;
(SAQs 8, 13–15, 19, 24)
(d) carbon, compared with its second-row neighbours; types of carbide; CO and CO_2; graphite;
(SAQs 16–18, 22)

7 Recognize compounds of boron, nitrogen, oxygen and the halogens that form Lewis acid-base complexes; and show how Lewis acids and bases react with weakly bonded compounds.
(SAQs 14, 18, 24)

8 Identify properties characteristic of the following compounds:
(a) oxides, ions, and oxy-acids of nitrogen, NO, NO_2, N_2O, N_2O_4, NO_2^+, HNO_2, HNO_3 (for these, relate their properties with the Lewis structures);
(SAQ 20)
(b) liquid ammonia, aqueous ammonia.
(SAQs 19, 24)

9 Classify the oxides, as they vary across the Periodic Table, in terms of their acid-base properties and structure (with information also from Units 6 and 7).
(SAQ 23)

10 Identify typical Lewis acid-base complexes formed by compounds of boron, nitrogen, oxygen, the halogens, etc.
(SAQs 15, 19, 24)

11 List properties of ammonia, water, and hydrogen fluoride which are consequences of hydrogen bonding, and give evidence for the electrostatic nature of the hydrogen bond in these compounds.
(SAQs 14, 22, 24)

2

12 Account for the weakness of the N—N and O—O single bonds compared with homonuclear bonds formed by the corresponding third-row atoms, and with bonds of N or O with H or C (atoms without lone pairs).
(SAQs 21, 22)

13 Interpret the characteristic properties of second-row elements and their compounds (listed in Objectives 6–9) in terms of the following principles, etc., from earlier Units:
(a)   the First Law of Thermodynamics; the use of Born-Haber or other thermodynamic cycles to analyse differences between the values of $\Delta H_m^\ominus$ for analogous reactions; standard enthalpies of reaction calculated from tables of thermodynamic data;
(b)   the increase in thermal stability of salts down Groups I and II when the solid product contains a smaller anion, as explained in terms of lattice energies;
(c)   the distinction between kinetic inertness (caused by slow rate of reaction) and thermodynamic stability;
(d)   differences in the coordination numbers of cations in Group I and II halides as accounted for by the relative sizes of ions;
(e)   valence shell repulsion theory, applied to predict shapes for simple molecules of typical elements;
(f)   the concepts of electronegativity or ion polarization, as they rationalize the different degrees of ionic character in the halides of different elements;
(g)   the periodicity of properties such as ionization energy, electron affinity, electronegativity, and covalent and ionic radius, as explained (qualitatively) in terms of energy shells of electrons in atoms, and of the effective nuclear charge.
(SAQs 9, 10, 12–17, 20, 22, 23, 25)

Table A

List of scientific terms, concepts and principles in Unit 9

Introduced in a previous Unit	Unit Section No.	Developed in this Unit	Page No.
	*S100**	catenation	42
electronic configuration	7.2.2	condensed oxyanions	29
electron spin	7.3	delocalized bonds	16
Hund's rule	7.4	diagonal relationship	25
Pauli's principle	7, App. 2	heteronuclear molecule	12
		homonuclear molecule	5
	S25-		
antibonding orbital	8.3	interstitial	50
aufbau procedure	8.2	isoelectronic	12
bond order	8.3	isotropic and anisotropic	20
bonding orbital	8.3	Lewis acid	27
Born—Haber cycle	5.8	Lewis base	27
covalent radius	8.6.4	lone pair electrons	9
effective atomic number	8.6.5	$p\pi$ bonds	9
electron affinity	5.8	π electron cloud	9
hybridization	8.4	polyanions	29
hydration	5	refractory	25
hydrogen bonding	8.8.1	solvation of ions	46
ionic radius	7.1.5	sp hybridization	14
non-bonding orbital	8.4.2	sp² hybridization	15
orbital overlap	8.3	sp³ hybridization	17
paramagnetism	8.1	three-centre bonds	23
π orbital	8.3.1		
σ orbital	8.3		
redox reactions	6.1.7		

* The Open University (1971) S100 *Science: A Foundation Course,* The Open University Press.

Introduction

In Unit 8, we studied a simple molecular orbital theory of chemical bonding which gives a satisfactory account of some of the molecules that prove troublesome in simple Lewis theory. For very simple molecules like H_2, molecular orbital theory is quantitatively correct, and by using hybridization the theory can predict (or at least accommodate) the shapes of some simple polyatomic molecules. The molecules we discussed in Unit 8 contain only single bonds. In this Unit, we extend the theory to include the formation of double and triple bonds. To start with we examine some diatomic molecules formed by the elements of the second row or period of the Periodic Table, Li to Ne. We then consider some polyatomic molecules involving multiple bonds. Throughout, we use the same criteria of suitable energy and orbital overlap as used in Unit 8 to determine molecular orbital formation.

We then continue with the descriptive chemistry of the lighter elements, from lithium to oxygen across the second row.

We find, particularly in the middle Groups, that distinctive properties mark off the first member of the Group from its successors. Indeed, on the basis of chemical properties, boron, nitrogen and oxygen can hardly be said to belong with the other elements of their respective Groups.

These differences can be explained by the small sizes of the second-row atoms and ions.

9.1 Diatomic molecules of the second row

We can now begin to apply the methods we developed in the previous Unit to the elements of the second row. Because we are concerned with discrete molecules, we shall compare our results with those diatomic molecules which have been observed as gases: Li_2, B_2, C_2, N_2, O_2, F_2. Molecules made up of atoms of the same element are called *homonuclear*.

homonuclear molecule

Of course, the elements do not all exist under normal conditions as diatomic molecules: elemental lithium is a metallic solid, boron is a non-metallic solid and carbon exists as graphite or diamond. These solids are more stable than the discrete molecules at normal temperatures. In this Section, we shall concentrate on the stabilities of the diatomic molecules relative to the isolated atoms. This is what we shall compare with the theoretical predictions.

As you saw in Unit 8, we have to decide which molecular orbitals are formed and which are occupied in order to determine whether a molecule is stable with respect to its constituent atoms. To do this we use the rules that were outlined in Unit 8, Sections 8.3 and 8.4, relating to the symmetries and energies of the atomic orbitals:

molecular orbital

1   Only those atomic orbitals of the same symmetry with respect to rotation about the internuclear axis combine to produce molecular orbitals. The two $2p_x$ orbitals combine, but the combination of $2p_x$ and $2p_y$ does not generate a molecular orbital because there is no net overlap of these two orbitals.

2   We need to consider only the valence shell orbitals. This is an energetic condition which arises because atomic orbitals combine only when their energies are similar, e.g. in LiH the H 1s and Li 2s combine, but H 1s and Li 1s do not, as shown in Unit 8, Figure 17.

3   Electrons occupy the lowest energy orbitals consistent with the Pauli principle (a maximum of two electrons in each orbital) and Hund's rule (when degenerate orbitals are available there is a maximum number of unpaired electrons with parallel spins).

9.1.1 Li_2

Lithium has the electronic configuration $1s^2 2s$. Qualitatively the problem is simple: the formation of Li_2 is analogous to the formation of H_2. Because we need only consider the valence shell orbitals we can ignore the 1s orbitals. We can understand this in terms of orbital overlap. As two lithium atoms approach each other, it is the 2s orbitals that overlap. The 1s electrons are much closer to the nucleus than the 2s electrons (TV programme for Unit 8), so the 1s orbitals do not overlap to any great extent. That is, the 1s electrons effectively retain their atomic identity. So it is the 2s orbitals that we need to consider. By a linear combination of the 2s atomic orbitals, a bonding and an antibonding orbital are generated (Fig. 1). The top part of Figure 1 is analogous to the formation of $\sigma 1s$ and $\sigma^* 1s$ orbitals in H_2 (Unit 8).

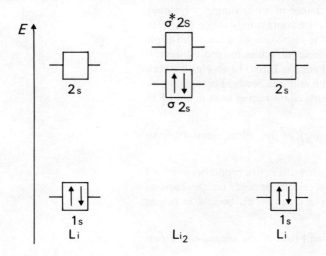

Figure 1   Energy level diagram for the formation of Li_2.

Does the bonding orbital have σ or π symmetry?

The molecular orbital is generated from two spherical atomic orbitals and therefore has σ symmetry.

The bond energies and bond lengths in H_2 and Li_2 are:

	H_2	Li_2
$\dfrac{D}{kJ\ mol^{-1}}$	436	105
$\dfrac{r}{pm}$	74	267

How do the predicted bond orders in these two molecules compare with the results?

The bond order is one in both molecules, but the difference in stability and bond lengths is remarkable. In the atoms the valence electrons are held much more tightly in hydrogen than in lithium (cf. the ionization energies in Unit 8). This difference in ionization energy is reflected in the calculated electron distribution. The radius of maximum charge density in the hydrogen 1s atomic orbital (53 pm) is considerably smaller than in the Li 2s atomic orbital (150 pm). The radius of the 1s atomic orbital in lithium is even smaller (20 pm) than that of the 1s orbital in hydrogen.

These differences confirm that when two lithium atoms approach each other it is the 2s orbitals that overlap. The 1s orbitals are essentially unaffected, which is why $\sigma 1s$ and $\sigma^* 1s$ molecular orbitals are not drawn in Figure 1. Compared with hydrogen then, overlap occurs at a relatively large distance between orbitals which are larger and consequently of lower electron density. This combination of distance and charge density is responsible for the weaker forces of attraction between the two lithium atoms. So, in terms of electron distribution, we can get a qualitative understanding of the bond characteristics in H_2 and Li_2.

9.1.2 Be_2

This molecule was not listed at the beginning of this Section. The electronic configuration of the beryllium atom is $1s^2 2s^2$. All the electrons are paired according to the Pauli principle. The molecular energy diagram is similar to that of Li_2 (see Fig. 1).

Can you see why Be_2 does not form?

The answer lies in the orbital occupancy. Be_2 has four valence shell electrons,

6

and they occupy the $\sigma 2s$ and $\sigma*2s$ orbitals, with two electrons in each molecular orbital. The net bond order is zero, and Be_2 is predicted to be unstable relative to the individual atoms. In fact, it has never been observed, and this negative result is a significant confirmation of the principles of molecular orbital theory.

9.1.3 B_2

Boron has the electronic configuration $1s^2 2s^2 2p$, so boron has three valence shell electrons. The molecule B_2 therefore has six valence electrons.

We now have to decide what energy level diagram is appropriate for the molecule. In Unit 8, Section 8.3.1 we considered the combination of excited atomic orbitals of hydrogen. The orbital pattern in more complex atoms is similar to that in hydrogen (Unit 8, Section 8.2), so we know which orbital combinations are possible according to symmetry arguments, i.e. which orbital combinations generate molecular orbitals. We also know that only orbitals of similar energies are suitable for combination. Figure 2 shows the energy sequence in the boron atom and suggests how the atomic orbitals should be combined.

As Figure 2 shows, the atomic orbitals are divided energetically into three groups: 1s, 2s and 2p. As we have seen with Li_2 and Be_2, the electrons in the 1s orbital tend to be relatively close to the nucleus, and again we need not consider these orbitals when we construct molecular orbitals.

We expect the 2s atomic orbitals to combine to produce one bonding orbital ($\sigma 2s$) and one antibonding orbital ($\sigma*2s$) exactly as in Li_2 and Be_2.

The six 2p orbitals of the two atoms are degenerate, so their energies give us no guidance in deciding how the molecular orbitals are formed. However, the 2p orbitals are not spherical but are directed along each of three axes. In Unit 8, Section 8.3.1, we considered how molecular orbitals are formed when two atoms approach each other along one of these axes, the x axis. We found that the symmetry of the atomic orbitals about that axis determined whether the orbitals overlapped in a manner suitable for molecular orbital formation.

Let us suppose that the two boron atoms approach along the x axis.

> Which atomic orbitals couple to satisfy the overlap condition? Figure 15 in Unit 8 will help you to answer this.

The two $2p_x$ orbitals combine to produce the $\sigma 2p$ and the $\sigma*2p$ molecular orbitals (Unit 8, Fig. 15a). The $2p_z$ orbitals produce $\pi 2p$ and $\pi*2p$ orbitals (Unit 8, Fig. 15c), and the $2p_y$ atomic orbitals also combine to give a $\pi 2p$ and a $\pi*2p$ orbital. Thus six atomic orbitals generate three bonding and three antibonding orbitals. Similarly, if the two atoms approach along the y or z axes, three bonding and three antibonding orbitals are produced.

The question now is: what is the energy sequence of these orbitals? Electrons in a bonding orbital stabilize the molecule, so by definition the bonding orbitals are those of lower energy.

Because an electron in the $\sigma 2p$ orbital is concentrated along the internuclear axis, whereas an electron in the $\pi 2p$ orbital is concentrated above and below this axis, the $\sigma 2p$ electron might be expected to attract the two nuclei more strongly. Now, an electron in a *bonding* orbital stabilizes the molecule to approximately the same extent as an electron in the corresponding *antibonding* orbital destabilizes the molecule, so we expect the six orbitals to increase in energy in the sequence

$$\sigma 2p < \pi 2p \text{ (two)} < \pi*2p \text{ (two)} < \sigma*2p$$

Figure 3 (overleaf) shows this energy sequence (sequence A) and the energy levels of the parent atomic orbitals.

A boron atom has three valence shell electrons, so there are six electrons to occupy the molecular orbitals. Applying the aufbau rules to Figure 3 we expect the $\sigma 2s$, $\sigma*2s$ and $\sigma 2p$ orbitals to be fully occupied (two electrons in each).

However, the electronic spectrum of B_2 shows that the two electrons in the highest energy orbitals are unpaired and have parallel spins! Since Hund's rule must be obeyed, these electrons only have parallel spins if the two degenerate

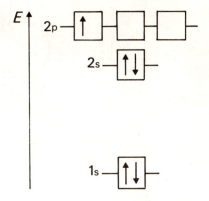

Figure 2   Energy levels in boron.

7

$\pi2p$ orbitals have lower energy than $\sigma2p$. The energy sequence (sequence B) must be that in Figure 4, which also shows the orbital occupancy in B_2.

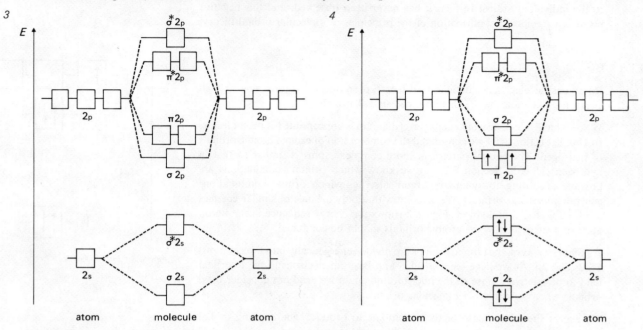

In fact the $\sigma2p$ and $\pi2p$ orbitals have very similar energies and the sequence in which they occur depends upon the internuclear distance of the molecule concerned. Molecular orbital calculations give the sequence of orbital energies. For shorter internuclear distances, as in B_2, C_2 and N_2, energy sequence B is correct. Energy sequence A applies to molecules with longer internuclear distances, such as O_2 and F_2. These calculations are confirmed by spectroscopic studies.

Figure 4 tells us the electronic state or configuration of B_2. It is $(\sigma2s)^2 (\sigma*2s)^2 (\pi2p)^2$, with the two π electrons unpaired.

> What property do you associate with unpaired electrons?

One consequence of the spin of an unpaired electron in an atom or molecule is manifest in the magnetic behaviour. The atom or molecule is paramagnetic, i.e. it is drawn into a magnetic field (Unit 8, Section 8.1). B_2 has two unpaired electrons with parallel spins (Fig. 4), so we expect the molecule to be paramagnetic.

There is a net excess of two bonding electrons in the molecule (the $\sigma2s$ and $\sigma*2s$ have a net bonding effect of approximately zero) and the bond order is therefore one. The bond dissociation energy is 289 k J mol^{-1} and bond length is 159 pm, values intermediate between those for H_2 and Li_2. Evidently there is a wide range of bond lengths and dissociation energies for single bonds, and the bond order alone tells us little about the bond characteristics, particularly when the bonding orbitals are derived from different types of atomic orbitals: $1s$ for H_2; $2s$ for Li_2; and $2p$ for B_2. The bond order in these instances is only a rough measure of bond strength. There is, however, a good correlation between theory (bond order) and experiment (bond length and energy) for bonds derived from similar atomic orbitals, as you will see shortly.

Figure 3 (left) Energy sequence A.

Figure 4 Energy sequence B; electrons in the molecule B_2.

9.1.4 C_2 and N_2

Having established the molecular orbital diagrams for the diatomic molecules from B_2 to F_2 across the Periodic Table, we can now examine whether the predicted bond orders and other properties are supported by experimental evidence.

Exercise 1

Figure 5 reproduces the energy level diagram that we need for C_2 and N_2. Using arrows to represent the valence electrons, fill in the boxes in Figure 5 and determine what bond orders are predicted for C_2 and N_2. Remember that only valence shell electrons are available for occupying the molecular orbitals.

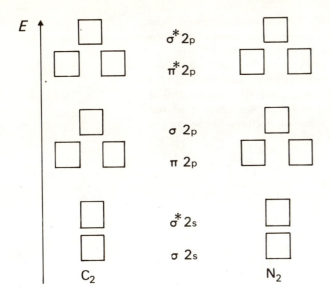

Figure 5 *Energy level diagrams for C_2 and N_2.*

Now compare your conclusions with the experimental data in Table 1.

Table 1   *Data for C_2 and N_2*

	C_2	N_2
dissociation energy $\dfrac{D}{\text{kJ mol}^{-1}}$	596	933
bond length $\dfrac{r}{\text{pm}}$	124	109.4
paramagnetic	no	no

Notice that the value for C_2 is not the value for any of the carbon-carbon bonds listed in Section 8 of the *Chemistry Data Book**. The values there are bond energy terms for carbon-carbon bonds in hydrocarbons.

The answer to Exercise 1 is on p. 55 after the SAQ answers and comments.

So far we have concentrated on the energies of the molecular orbitals and how they affect bond lengths and bond strengths. With diatomic molecules there are no problems of geometry, but the predicted distribution of electrons is interesting and we can compare this prediction with the predictions of Lewis' theory. For nitrogen the octet rule requires a triple bond, and the remaining two pairs of valence shell electrons are the lone pairs.

$$\overset{\times}{\underset{\times}{\times}}\text{N}\,\vdots\,\text{N}\,\vdots$$

In molecular orbital theory we can regard the $\sigma2s$ and $\sigma*2s$ electrons as corresponding to the lone pair electrons. The $\sigma2p$ electrons are concentrated along the internuclear axis but the $\pi2p$ orbitals, which are generated by the combinations $2p_y + 2p_y$ and $2p_z + 2p_z$ (Unit 8, Fig. 15c), concentrate electrons around the internuclear axis, but not along it. The overall effect of these π orbitals is to produce a cylinder of electron cloud surrounding the internuclear axis (Fig. 6).

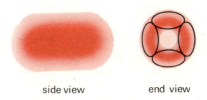

side view   end view

Figure 6   *π electron clouds in N_2.*

lone pair electrons

$p\pi$ bonds
π electron cloud

9.1.5 O_2 and F_2

Figure 3 gives the correct energy sequence of orbitals for the molecules O_2 and F_2. O_2 has 12 valence shell electrons and F_2 has 14. Applying the aufbau rules, and using Hund's rule, we get the electron distribution for O_2 shown in Figure 7.

The molecular orbital occupancy in F_2 is shown in Figure 7. The fluorine molecule has a net excess of two bonding electrons (a single bond) and the electrons are all paired.

* The Open University (1973) *S24-/S25- The Open University Chemistry Data Book*, The Open University Press.

O_2 has eight electrons in bonding orbitals and four in antibonding orbitals. Thus O_2 has a net excess of four bonding electrons, a double bond. Notice that the two electrons in the degenerate $\pi*2p$ orbitals are unpaired. Oxygen molecules are therefore predicted to be paramagnetic; this property can be illustrated simply. As we show in the TV programme, liquid oxygen is drawn into the field of a strong magnet. This agreement between experimental observation and theoretical expectation strongly confirms the molecular orbital approach. There is no way that simple Lewis theory can account for unpaired electrons in the oxygen molecule.

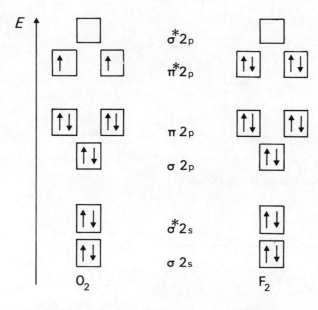

Figure 7 Orbital occupancies in O_2 and F_2 (sequence A).

Recently, calculations of molecular orbitals have come within about 1 per cent of the experimental bond lengths of some diatomic molecules. The sort of agreement achieved is shown in Table 2, which also includes the bond dissociation energies of O_2 and F_2 for comparison with the bond orders given by Figure 7.

Table 2 Data for O_2 and F_2

	O_2	F_2
bond order	2	1
bond dissociation energy $\dfrac{D}{kJ\ mol^{-1}}$	498	158
bond length (experimental) $\dfrac{r}{pm}$	120.74	143.5
bond length (calculated) $\dfrac{r}{pm}$	120	142
paramagnetic	yes	no

SAQ 1 (Objective 2)   What are the electron configurations of the two ions N_2^+ and O_2^-? What are the bond orders predicted by molecular orbital theory for these two ions?

Assume that the energy sequences for these ions are the same as for the corresponding molecules.

9.1.6   Summary of the bonding in second-row elements

In this Section we have concentrated on determining the bond orders of the diatomic molecules formed by the elements from lithium to fluorine in the Periodic Table. At first glance the predictions of molecular orbital theory look promising; the molecule Be_2, which has a bond order of zero, has not been observed, but all the other diatomics have been observed, if only at high temperatures (e.g. Li_2 and C_2). As we pointed out earlier, there is not an exact relation between bond order and the measurable properties of bonds such as length and strength. However, theory and experiment can be compared neatly by representing the results in graphical form.

Plot the following properties of the diatomic molecules against the number of valence electrons in the molecule (e.g. two in Li, four in Be_2, etc.).

Properties: bond order
bond dissociation energy
bond length.

You do not have the experimental values for Be_2 since Be_2 is not known to exist. You will have to assume that it is not stable with respect to two Be atoms, in which case

$$D(Be - Be) = 0$$

The correlation between bond order and the experimental properties is quite striking, as is the prediction of paramagnetic properties. Molecular orbital theory evidently provides a very satisfactory explanation of the bonding in these molecules.

SAQ 2 (Objective 2)   Using the energy level diagram in Figure 3, what can you say about the existence of the molecule Ne_2?

9.1.7   The singularity of the second-row elements, lithium to neon

These elements are characterized by one feature; in the gas phase at least, they tend to form diatomic molecules and several of these are multiply bonded. Admittedly lithium, boron and carbon only do so at elevated temperatures, but the gaseous diatomic molecule is the stable form of oxygen, nitrogen and fluorine. In molecular orbital terms, this property of multiple bond formation implies an ease of formation of π bonds through overlap of the p orbitals.

Chemical periodicity leads us to expect gaseous diatomic molecules such as P_2 and S_2 for elements of the next period. However, phosphorus and sulphur both exist as solids under normal conditions, although there is spectroscopic evidence that S_2 exists at elevated temperatures. The solids are characterized by single σ bonds. In this respect, the elements of the second row are unrepresentative. Evidently the elements of the second row tend to form multiple bonds through pπ orbital formation more readily than the elements of the third and subsequent rows. Several attempts have been made to rationalize this observation.

Let us examine the available data on the strengths of the relevant bonds. Table 3 lists the bond lengths and bond energy terms (E) for some single and multiple bonds. The bond-energy term is a measure of the strength of the bond (Unit 8, Appendix 1).

Table 3   Bond lengths and bond energy terms

	Second-row elements		Third-row elements	
bond	N—N	O—O	P—P	S—S
$\dfrac{E}{kJ\ mol^{-1}}$	159	142	209	264
$\dfrac{r}{pm}$	147	149	221	206
bond	N≡N	O=O	P≡P	S=S
$\dfrac{E}{kJ\ mol^{-1}}$	933	498	524	428
$\dfrac{r}{pm}$	110	121	190	189

The data for the multiple bonds are obtained from studies of the diatomic molecules. The single bond data are estimated by studying compounds like

11

hydrazine (N_2H_4) and its phosphorus analogue (P_2H_4), hydrogen peroxide (H_2O_2) and elemental sulphur (S_8) in which the atoms are joined by single bonds. A count of the valence electrons in these compounds reveals that the atoms have lone pairs of electrons, e.g. one lone pair on each oxygen atom in H_2O_2.

Some correlations are obvious and expected. Multiple bonds are shorter and stronger than single bonds between the same atoms. However, there are striking differences between the data for second- and third-row elements. In the second-row elements, the atomic orbitals are smaller than in the third-row elements. The orbital overlap necessary for the formation of a bond, therefore, occurs at shorter distances in the second-row elements. This applies to both single and multiple bonds and is reflected in the bond lengths.

If we concentrate on the single bonds (the σ bonds), we see that they are weaker in the second-row elements than in third-row elements. At the shorter bond distances in the second-row elements, the lone pairs of electrons are forced closer together and the consequent repulsion between these lone pairs is greater. The result is a weaker bond.

On the other hand, the π bonds are stronger in the second-row elements than in the third-row elements. A very rough measure of the strengths of the π bonds can be obtained from the difference between the multiple and single bond strengths in Table 3. (It is rough because the strength of multiple and single bonds depends on the bond length and this changes from single to multiple bonds.) Although the p orbitals overlap at longer distances in the third-row elements, it is argued that the extent of overlap of p orbitals is less than in the second row.

Now summarize what you consider to be the main points of this Section and compare your summary with that given below.

1   Molecular orbital theory describes the bonding in diatomic molecules that are formed by the elements of the second row of the Periodic Table. Apart from beryllium and neon, the diatomic molecules of these elements are predicted to be stable relative to the gaseous atoms, although under normal conditions more stable solid forms of lithium, boron and carbon exist.

2   The appropriate energy level diagram for each molecule is determined by calculation, and these energy level diagrams are generally supported by spectroscopic evidence.

3   Molecular orbital occupancies in these diatomic molecules predict bond orders, which correspond to experimental bond energies, and paramagnetic properties.

4   The formation of strong π bonds through the overlapping of p orbitals in the second-row elements contrasts with the bonding in elements of other rows of the Periodic Table.

9.2   Heteronuclear diatomic molecules

In Section 8.4 of Unit 8, you have already met two molecules of this type, lithium hydride and hydrogen fluoride. In both of these molecules, atomic orbitals combine to form molecular orbitals according to suitable conditions of symmetry and energy. For example, in LiH the σ orbital is formed from H 1s and Li 2s, i.e. the valence shell orbitals combine to form molecular orbitals. When molecules form between atoms of the second row of the Periodic Table (Li to Ne), the electrons occupy orbitals that are formed from the 2s and 2p atomic orbitals. The picture is much the same as in the homonuclear diatomic molecules. The sequence of energy levels or orbitals is given either by Figure 3 or by Figure 4. We shall consider two molecules, CO and NO, both of which have energy sequence B (see Fig. 4 on p. 8).

CO has ten valence shell electrons (C, $2s^2 2p^2$, and O, $2s^2 2p^4$). This is the same number as in N_2, so we say that CO is *isoelectronic* with N_2. With the same energy level sequence for the two molecules, we expect CO to have a bond order of three and, if we distribute the ten valence electrons among the orbitals in Figure 8, we get the electronic configuration $(\sigma 2s)^2 (\sigma^* 2s)^2 (\pi 2p)^4 (\sigma 2p)^2$. As with nitrogen, the bond is very strong and also short.

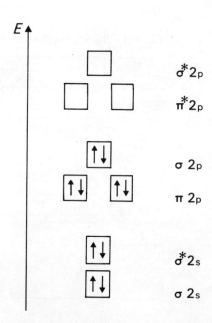

Figure 8   Energy level diagram for the heteronuclear molecule CO.

12

Table 4   Data for CO

	Experimental	Calculated
bond dissociation energy $\dfrac{D_m(\text{C-O})}{\text{kJ mol}^{-1}}$	753	1090
total electronic energy* $\dfrac{E}{\text{kJ mol}^{-1}}$	296 100	297 700
bond length $\dfrac{r}{\text{pm}}$	113	1 10

*Sum of the energies of all the electrons (valence and inner shell, i.e. 1s electrons) in the molecule.

Carbon monoxide is a well characterized compound. Spectroscopic measurements give the bond dissociation energy and bond length. Table 4 compares the experimental data with the results of a recent calculation in which electron correlation energies were considered. Notice that the bond length and total electronic energy are calculated with high accuracy, but the dissociation energy involves a large discrepancy. As we stated in Unit 8, molecular orbital theory is still a long way from calculating accurate bond dissociation energies for anything but the simplest molecules, even when very large computers are available.

At the end of Unit 7, you saw that one exception to the electron pair bond as the essential element of chemical structure is exhibited by nitric oxide. The NO molecule contains eleven valence shell electrons, yet NO is a colourless gas showing little tendency to dimerize at normal temperatures and pressures.

The bond in NO is found by experiment to be very short (115 pm) and very strong (678 k J mol^{-1}). Compare this value with the double bond in O_2 (498 kJ mol^{-1}).

> What is the electronic configuration of NO and what bond order is predicted? (Figure 8 gives the sequence of energy levels.)

The net excess of bonding electrons is five, giving a bond order of $2\frac{1}{2}$. The electronic configuration of NO is $(\sigma 2s)^2 \ (\sigma^*2s)^2 \ (\pi 2p)^4 \ (\sigma 2p)^2 \ (\pi^*2p)^1$. This emphasizes a point made in Unit 8: in contrast to the Lewis theory of electron-pair bonding, the essential unit of bonding in molecular orbital theory is the single electron in a bonding molecular orbital.

Now summarize what you consider to be the main points of this Section and compare your summary with that given below.

1   Molecular orbital theory accounts for the high strength of the bond in carbon monoxide (a triple bond).

2   The odd electron structure of nitric oxide is explained and the bond order found to be $2\frac{1}{2}$.

> SAQ 3 (Objectives 2,3)   The following spectroscopic measurements have been made on the molecule NO and ion NO$^+$; $r_{NO} = 115$ pm; $r_{NO^+} = 106$ pm. Are these values consistent with the predictions of molecular orbital theory?

9.3   Hybrid orbitals and multiple bonding in polyatomic molecules

Having studied how the geometries of molecules can be accounted for by the shapes of simple molecular orbitals and hybrid orbitals, and having seen how multiple-bond orders of two and three arise in molecular orbital theory, we can now turn to examine the shapes and bond orders in some polyatomic molecules. In Unit 8 we deferred the discussion of the bonding in beryllium hydride and boron hydride. We now take up the problem of the shape of BeH_2.

Beryllium

Beryllium has the electronic configuration $1s^2 2s^2$. With two valence shell electrons, a simple linear molecule BeH_2 is predicted by electron pair repulsion theory. Since the 2s valence electrons in beryllium are paired, electron sharing requires an unpairing by the excitation of an electron from 2s to 2p. With the electronic configuration $1s^2 2s 2p$, the Be atom is expected to form two bonds, and the question of bond angle arises. The s orbital is spherical and the p orbital has directional properties, so, as with methane (Unit 8, Section 8.4) it is not immediately apparent what shape is predicted for BeH_2. Unfortunately, experiments provide only indirect information. Discrete molecules of BeH_2 have not been observed, although spectroscopic measurements show that the analogous molecule, $BeCl_2$*, is linear with the two Be—Cl bonds of equal length. Obviously, this shape is not easily accounted for using simple atomic orbitals.

electron promotion

One way around this problem is to extend the concept of hybridization, which we introduced to account for the tetrahedral shape of methane. For $BeCl_2$, the Be atom requires two equivalent atomic orbitals directed at 180° from each other.

This involves making a linear combination of the 2s and 2p orbitals of Be before bond formation. The two resulting orbitals are linear and are called *sp hybrid orbitals*. Figure 9 shows the 2s and 2p orbitals on the left and the two sp hybrid orbitals on the right.

sp hybridization

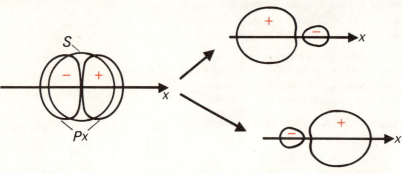

two sp hybrid orbitals

Figure 9   The formation of sp hybrid orbitals.

The linear geometry of $BeCl_2$ can now be explained by combining the sp hybrid orbitals of Be with suitable orbitals of chlorine, e.g. 3p.

Clearly, overlapping of the sp hybrid orbitals with the 1s orbitals of hydrogen also leads to a linear molecule. The combination of the sp and 1s orbitals gives two bonding and two antibonding molecular orbitals. In fact, the linear geometry gives the maximum orbital overlap and minimum energy for BeH_2.

This two-stage process–hybridization followed by linear combination with the hydrogen orbitals–is combined in a single operation in the calculation, but it is convenient for us to think of it in two steps as an aid to understanding the geometry. The molecule BeH_2 is particularly interesting from the point of view of the theoretical chemist. Because it has not yet been observed as discrete molecules, and because theory predicts it to be stable with respect to Be and H, it offers the opportunity to make theoretical predictions of other molecular parameters, e.g. bond lengths, force constants and bond dissociation energies.

Boron

With the electronic structure $2s^2 2p$, boron might be expected to form compounds like BH_3, BF_3 and BCl_3.

What shape is predicted for BF_3 or BH_3 by electron pair repulsion theory?

Both molecules have three valence electron pairs and are predicted to be trigonal planar. Indeed the BF_3 molecule does have this shape. But the boron atom has only one unpaired valence electron, the 2p, and according to molecular orbital theory we might expect the molecules BH or BF to be formed.

---

* This is observed in the vapour at 1 000 K. The structure of $(BeCl_2)_n$ at room temperature is described in Section 9.6.

However, boron forms a series of hydrides with unusual structures (we examine the bonding in the simplest of these B_2H_6 in Appendix 2). The stable fluoride is the planar molecule BF_3, which has three equivalent BF bonds separated by angles of 120°.

$$\begin{array}{c} F \\ | \\ F \diagdown \overset{\displaystyle B}{} \diagup F \end{array}$$

This capacity to form three equivalent bonds presents us with a similar dilemma to those we encountered with methane (Unit 8, Section 8.4.3) and $BeCl_2$. The geometry of methane can be accommodated only by the formation of sp^3 hybrid orbitals or an equivalent procedure. Although detailed calculations agree that the tetrahedral structure corresponds to the most stable arrangement, this structure is not easily predicted by simple arguments.

In boron, promotion or excitation of an electron from 2s to 2p requires similar energy to the same process in carbon, and gives boron an increased bonding capacity due to three unpaired electrons. With the electronic configuration 2s $2p_x \, 2p_y$ produced by promotion, boron has one electron in the spherical s orbital and two electrons in perpendicular p orbitals. It is still not apparent why the three bonds in BF_3 are equivalent and at 120° to each other.

The problem is again reminiscent of that of methane: a number of atomic orbitals of a central atom, having unequal energies, produce the same number of equivalent bonds. Again hybridization provides an answer to the problem. This time the three atomic orbitals 2s, $2p_x$ and $2p_y$ are combined or hybridized to give three equivalent sp^2 orbitals which are planar and directed trigonally (at 120° to each other) as shown in Figure 10. Each hybrid orbital on the right-hand side of the figure consists of one lobe where ψ is positive and a smaller lobe where ψ is negative.

sp² hybridization

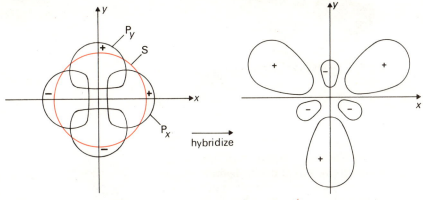

Figure 10   The formation of sp² hybrid orbitals.

If three fluorine atoms are arranged trigonally around the boron atom, each positive lobe of the sp^2 orbitals of boron can overlap with a suitable orbital of fluorine to give bonding and antibonding orbitals. In its valence shell fluorine has one half-filled 2p orbital, so we can assume that this orbital overlaps with an sp^2 orbital. The result is that three sp^2 boron orbitals combine with a 2p orbital of each fluorine atom to give six molecular orbitals, three bonding and three antibonding. Boron has three valence shell electrons and fluorine has one electron in the half-filled 2p orbital used in bond formation–a total of six electrons which occupy the three lowest energy orbitals, the three bonding orbitals, in BF_3.

Would you describe these orbitals as σ or π?

Again hybridization is merely a step in the calculation of molecular orbitals from atomic orbitals, but it does preserve the identity of certain orbitals with chemical bonds, i.e. the idea of a pair of electrons localized in the region between two nuclei is retained through the localized orbital.

Now we have said that BF_3 is a stable compound but that BH_3 has not been observed. You might argue that the combination of sp^2 hybrid orbitals with the 2s orbitals of hydrogen could provide the basis of a stable molecule, BH_3. This argument would be quite consistent with the experimental facts; the reason that BH_3 is not observed is that B_2H_6 is a much more stable molecule and always forms at the expense of BH_3. So why is BF_3 stable? The BF_3 molecule is stabilized by bonding that we have not yet considered.

The sp^2 hybrid orbitals are symmetrical with respect to rotation about the bonding axes as is the overlapping 2p orbital of each fluorine, so the bonds are σ bonds.

Which orbital of boron have we not considered in the formation of the sp^2 hybrid orbitals?

In boron, this is the $2p_z$ orbital which is perpendicular to the plane of the sp^2 orbitals and is unoccupied. But the parallel orbitals ($2p_z$) of the fluorine atoms are each fully occupied, and furthermore the orbitals of fluorine overlap with the $2p_z$ orbital of boron. So we have four overlapping atomic orbitals which can combine to produce four molecular orbitals. Of these, one is bonding, one is antibonding, and the other two make no contribution to the stability of the molecule. They are non-bonding orbitals with energies between those of the bonding and anti-bonding orbitals. With six electrons available (two from each fluorine $2p_z$ orbital), the three lowest energy molecular orbitals are occupied: the bonding and two non-bonding orbitals. The bonding orbital is of most interest to us and its formation is shown in Figure 11.

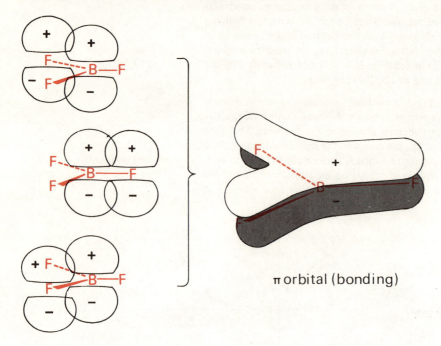

π orbital (bonding)

Figure 11   The formation of the delocalized π bonding orbital in BF_3.

The left-hand side of the Figure shows the overlap between B $2p_z$ and the $2p_z$ orbitals of each of the F atoms. The right-hand side shows the molecular orbital which has three arms and, unlike the σ orbitals formed from the sp^2 hybrids, it extends over all four atoms. Viewed from a position above the plane of the molecule, the trefoil shape of this orbital is apparent (Fig. 12). This diagram shows that the orbital is not localized between two particular nuclei; it is called a *delocalized orbital*.

Notice that the $2p_z$ atomic orbitals have the same symmetry with respect to the plane of the molecule; a positive lobe above the plane and a negative lobe below. The molecular orbital must have the same symmetry making the plane of the molecule a node. Figure 11 shows the two lobes ($+$ and $-$) of the molecular orbital sandwiching the atoms. With a nodal plane through the molecule this orbital therefore has π symmetry.

The maximum overlap of the $2p_z$ orbitals of B and F occurs when the molecule is planar, so the orbital makes its greatest contribution to the molecule's stability when this condition applies. This point is taken up in the TV programme for Unit 10 where we compare the infra-red spectrum of BF_3 with the spectrum of BF_3—NH_3. In the compound BF_3—NH_3 the $2p_z$ orbital of boron is used to form sp^3 hybrid orbitals, and the π orbital we have just considered does not exist. Consequently, the BF bonds are observed to be longer and weaker in BF_3—NH_3 than in BF_3.

Can you see why no π orbitals are formed in BH_3?

In Unit 7 you met two other trigonal planar structures, the ions NO_3^- and CO_3^{2-}. To explain these structures resonance hybrids were drawn.

delocalized bond

In hydrogen the $2p_z$ orbitals are unoccupied, so that there are no electrons available to occupy the π orbitals.

16

The bonding in these structures is analogous to that in BF_3. The ions are iso-electronic with BF_3; all three species possess twenty-four valence shell electrons. The orbitals of interest to us are first the 2s, $2p_x$ and $2p_y$ orbitals of the central atom. sp^2 hybrids of these lead to the trigonal planar structures. Secondly the occupied $2p_z$ orbital of the central atom overlaps with the parallel orbitals of the oxygen atoms to produce a π orbital which extends over the whole ion and reinforces its stability and planarity (compare with BF_3, Fig. 12).

Figure 12   Bird's eye view of the π bonding orbital in BF₃.

Carbon

Finally we consider some compounds of carbon that involve multiple bonding. In Unit 8, we saw that carbon forms four equivalent σ bonds in methane by sp^3 hybridization. We also discussed bonding in C_2, although this molecule is unstable with respect to the solid forms of carbon. Apart from the usual tetra-hedral structure which dominates its chemistry, carbon exhibits two other geometries: trigonal planar as in ethylene and linear as in acetylene. In addition to being planar, ethylene is also rigid; the two trigonal groups are not free to rotate about the double bond.

> What two aspects of molecular orbital theory can explain the geometry and rigidity of ethylene?

Just as carbon can be sp^3 hybridized as in methane, so sp^2 hybridization is also possible (the same s→p excitation is a necessary step). Overlap of one of the sp^2 hybrid orbitals of each carbon forms a σ bond between the two carbon atoms while each of the other two sp^2 hybrid orbitals overlaps with a 1s orbital of hydrogen to give σ bonds between carbon and hydrogen. These bonds are shown as the full and dotted red lines in Figure 13. The carbon and hydrogen atoms all lie in one plane. This leaves each carbon with one valence shell orbital, the out-of-plane 2p orbital which contains one electron. Linear combination of these two overlapping orbitals produces a π orbital which consists of an electron cloud above and below the σ bond (Fig. 13).

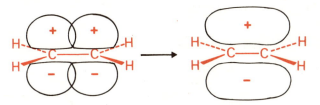

Figure 13   The formation of the π orbital in ethylene.

The two electrons occupy this orbital, which is of higher energy than the σ orbitals and also geometrically more accessible to approaching molecules. As a result the chemical behaviour of ethylene is determined largely by the π orbitals.

> What is the effect of rotating one CH_2 group so that it is perpendicular to the other?

For the unhybridized orbitals this is equivalent to the combination $2p_x + 2p_y$ (Unit 8, Fig. 15). There is zero net overlap of the 2p orbitals. Now you can see why the molecule is rigid in contrast to CH_3-CH_3. Rotation of the two CH_2 groups through 90° destroys the π overlap of the p orbitals shown in Figure 13.

In acetylene the linear geometry suggests sp hybridization as in the formation of BeH_2. Again, overlap of one hybrid orbital of each carbon produces a σ bond between the two atoms, and the atoms HCCH all lie in a straight line. This leaves each carbon with two perpendicular 2p orbitals each containing one electron, the $2p_y$ and $2p_z$ orbitals. The $2p_y$ orbitals of the two carbon atoms overlap in the same way as in ethylene to form a π orbital, which is fully occupied. Similarly, the $2p_z$ orbitals form a π orbital. The resulting orbitals form a cylindrical electron cloud surrounding the σ bond, exactly as in the N_2 molecule (Fig. 6). Once more the π bonding orbitals are the highest energy orbitals and this, coupled with the associated electron distribution, determines the chemical activity of triply-bonded carbon compounds.

At this stage it would be useful for you to note the main points of this Section and then do SAQs 4 to 7.

SAQ 4 (Objectives 1, 4 and 5) Ozone, O_3 is a bent molecule with a bond angle of 117°. The bond length, 128 pm, is intermediate between the O—O single bond length (149 pm in H—O—O—H, hydrogen peroxide) and the O=O double bond length (121 pm in O_2).

What aspects of molecular orbital theory account for the geometry and the bond length?

SAQ 5 (Objective 4) In the allene molecule, CH_2=C=CH_2, the central carbon atom is sp hybridized and the other two carbon atoms are sp^2 hybridized. Both carbon–carbon bonds are double bonds involving pπ orbitals.

What values would you predict for the C—C—C bond angle and for the H—C—H bond angle?

SAQ 6 (Objectives 4 and 5) How would you expect the two CH_2 groups in allene (SAQ 5) to be aligned with respect to each other?

SAQ 7 (Objectives 4 and 5) Benzene, C_6H_6, is a planar molecule with C—C—C angles of 120° and an observed bond length that suggests a bond order between one and two. This can be accounted for by resonance structures (see Unit 7, SAQ 17).

How can molecular orbital theory account for the bond angles and bond lengths in benzene?

SAQ 8 (Objectives 4 and 5) Figure 11 will help you to answer this.

In the molecular orbital formulation of the BF_3 molecule, what is the hybridization of the halogen? How are its four orbitals used?

On the resonance formulation of this molecule: write down canonical forms. What is the theoretical estimate, from these formulations, of the bond order in the B-F bonds. What is the experimental evidence that supports this bond order?

In your summary of the main points you probably noted that we use the idea of hybridization to account for three geometries:

1   linear   sp hybridization   e.g. $BeCl_2$, (BeH_2) and allene

2   trigonal sp^2 hybridization   e.g. BF_3, C_2H_4 and benzene

3   tetrahedral sp^3 hybridization   e.g. CH_4.

So far, in our discussion of chemical bonding in Units 8 and 9 we have confined our attention to relatively simple molecules of the type for which wave functions can be calculated.

You might have noticed that in our discussion in this Section and in the SAQs, we have begun to extend the scope to more complex molecules, particularly those of carbon, e.g. C_2H_4, C_3H_4 and benzene.

The concepts of localized hybrid orbitals (sp, sp^2 and sp^3) and delocalized π orbitals allow us to describe qualitatively the bonding in carbon compounds, including large organic molecules.

S100, Unit 8, describes some inorganic chemistry of carbon, including its allotropy. The structure of graphite is given again in Figure 14. This shows that the carbon–carbon distance between the layers (335 pm) is double the van der Waals radius of carbon (the non-bonded radius, as defined in Unit 8, Section 8.6.4).

In the regular hexagons of the graphite sheet (Fig. 14), the carbon–carbon distance is 142 pm. We can use this to estimate the bond order by comparing it

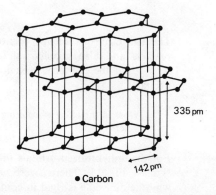

335 pm

142 pm

● Carbon

Figure 14 The structure of graphite.

with the distances for standard single, double, and triple bonds, as in Table 5 and Figure 15.

Table 5   Bond length and bond order in carbon–carbon bonds

Bond	Order	Length/pm	Example
C—C	1	154	ethane, diamond
C⚌C	about $1\frac{1}{3}$	142	graphite
C⚌C	about $1\frac{1}{2}$	139	benzene (*SAQ 6*)
C=C	2	133	ethylene
C≡C	3	120	acetylene

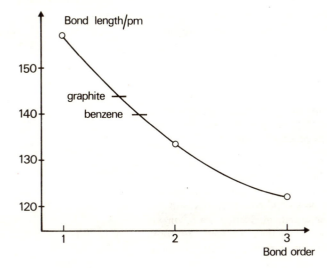

Figure 15   Bond lengths and bond orders of carbon-carbon bonds.

In graphite, as in benzene, the carbon–carbon bond is intermediate between a single and a double bond. Graphite, in effect, represents a series of layers which consist of an infinite extension of the arrangement found in aromatic compounds, or arenes (S24*, Unit 8) in which benzene rings are fused together, as shown in Figure 16.

We can account for the bond order in the graphite sheet in terms of the number of σ and π electrons.

> Of the four valence electrons, how many are used in σ bonding and how many in π?

To account for the trigonal arrangement of bonds about a carbon atom, we say that the carbon atoms are sp^2 hybridized. Overlap of these sp^2 orbitals between carbon atoms generates three σ orbitals between each carbon atom and its neighbours.

Three of the carbon's four valence electrons occupy these σ orbitals.

The remaining valence electron is in the $2p_z$ atomic orbital which is the one perpendicular to the plane of the graphite sheet. These $2p_z$ orbitals overlap in a similar way to those described for benzene (SAQ 6) to produce a number of π orbitals which extend over the whole graphite sheet; these are indeed multi-centre orbitals! The carbon atoms each supply one electron to occupy these π orbitals. Now, each carbon atom contributes one electron to the π orbitals and there are three bonds to carbon. Since a bond connects two carbon atoms the π bond order is about $\frac{1}{3}$, making a total bond order of about $1\frac{1}{3}$.

Figure 15 shows that the bond order, by interpolation, is about 1.5. In benzene, there is again one π electron for each carbon, but only two carbon ligands, so the π bond order is then $\frac{1}{2}$ on this model.

Figure 16   Valence bonds in the graphite layer.

multi-centre orbitals

* The Open University (1973) S24- *An Introduction to the Chemistry of Carbon Compounds*, The Open University Press.

We can interpret the physical and chemical properties of graphite in terms of its σ and π bonding: such properties as its electrical conductivity, lubricity, opacity, lustre, relatively low density, insolubility, inertness to chemical attack, and *anisotropy**. Graphite is unreactive and insoluble because the sheets are giant molecules held together by strong covalent bonds. Since there are no σ or π bonds to hold adjacent sheets together, they are relatively far apart (which accounts for the low density) and they readily slide over each other (which explains the lubricity).

isotropy and anisotropy

The giant π orbital resembles a two-dimensional metal; we shall return to this in Unit 10.

To summarize, we concluded this Section by showing how the bonding throughout a sheet of graphite (which is vast on the atomic scale) can be described by sp² hybridization and delocalized π orbitals. The description is consistent with the bond length and bond order and, although beyond the scope of this Course, also allows us to interpret many physical and chemical properties of graphite.

delocalized π orbitals

Study comment for Sections 9.4 to 9.11

In the remaining Sections of this Unit we explore some descriptive chemistry of the elements of the second row, from lithium to oxygen; fluorine will come into some of our generalizations but belongs in general in Unit 11.

As we go, we shall see how this chemistry is illuminated by, and illuminates, the concepts of structure and bonding and the thermodynamics that we have developed; you will find it convenient to have Units 5–8 to hand, for reference. In particular, we shall compare and contrast the second-row elements with the subsequent members of their groups. We remarked, in Section 9.1.6, on an important singularity of the second-row elements, their tendency to form multiple-bonded structures. We shall follow this through the chemistry of these elements, and sum up in Section 9.11.

Sections 9.4–9.11 contain a good deal of factual material. Frequently this material illustrates important principles, and these principles should help you to remember the examples. But if you are wondering whether you should try to memorize particular items of descriptive chemistry, refer to the Objectives for help.

Chapters 1, 3 and 4 of Johnson** contain background reading for this Unit, but finish the Unit first, so that you can recognize as black-page material anything in Johnson that is not treated in the white pages of Unit 9. Comments in Johnson on elements of the third and later rows anticipate Unit 10. We shall study acid-base properties, including the Brønsted-Lowry system, in Unit 11, Appendix 1 (White). Electrode potentials E^0 are black-page material, and so are the halides and ring compounds of nitrogen.

9.4 Chemistry of the elements of the second row

Many second-row elements are remarkably different from subsequent members of their Groups. Look at the Periodic Table, in Figure 22 of Unit 8. Only in Groups I, VII and VIII (the alkali metals, the halogens, and the noble gases) is there a name which covers all members of the group. Beryllium is not counted as an alkaline earth metal; its earth*** is not alkaline, being insoluble in water, and it forms almost no ionic compounds. In the middle Groups, there is no

---

* An isotropic substance is one that has the same properties regardless of direction in the structure. The word comes from the Greek words for equal and turn.

Graphite is *ani*sotropic, because of its layer structure. Thus, for example, if the electrical conductivity is measured in a direction parallel to the layers, it is about 10^5 times the conductivity measured perpendicular to the layers. This is because the π electrons can move freely within the layers, but not from one layer to the next.

** R. C. Johnson (1966) *Introductory Descriptive Chemistry*. Benjamin.

*** 'Earth' is the old name for a mineral oxide.

group name at all, since the elements are so assorted. Boron is a non-metal in a group of metals. Carbon differs greatly from silicon: indeed, boron is more like silicon in its chemistry than carbon is. Nitrogen and oxygen, gases in the elementary state, are unlike phosphorus, sulphur, and the rest.

> What generalization can you make about the covalent and ionic radii of the second-row atoms, compared with those of corresponding atoms in later rows (see Unit 8, Figs. 27, 28 and 32)? Note down the corresponding relationships of the ionization energies (Unit 8, Fig. 25). What conclusions can you draw about the strengths of binding of the second-row valence electrons, relative to those of subsequent rows?

The covalent and ionic radii increase down the Groups, but the increase from the second row to the third is much greater than the subsequent increments in Groups III to VII.

The atoms of the second-row elements, particularly the middle and later ones in the row, are significantly smaller than the corresponding ones in subsequent rows. Similarly, the ionization energy of the second-row atoms is significantly higher than in later rows. Thus, in these second-row elements the valence electrons are held relatively strongly.

> Can you relate this strong binding of the second-row valence electrons to the effective nuclear charge they experience (Unit 8, Section 8.6.5)? How do the electronegativities of these atoms fit into this picture?

The force holding the valence electrons increases with the effective nuclear charge $Z_{eff}e$, and decreases with their mean radius r. Figure 31 of Unit 8 shows that Z_{eff} is lower for the second-row elements than for the corresponding ones in later rows.

The important factors are the small radius and the small number of inner electrons, only two. The ionization energies are high because the valence electron can approach quite close to the nuclear charge (as Figure 2 of Unit 8 shows) and because repulsion by the inner electrons is small.

On the other hand, the second-row elements have relatively high electronegativities, reflecting the high ionization energies and small radii.

Now we shall see how these atomic properties are reflected in the chemical and physical properties of the elements of the second row.

9.5 Lithium

Lithium is a rare element, but is widely distributed in rocks as sodium and potassium are. Like them, it was first made by Davy (Unit 3, Appendix 1) by electrolysis of the moist oxide and is now made by electrolysis of the fused chloride.

In physical and chemical properties, lithium of all the second row elements (except neon) is the one that most resembles its group.

> Can this be said also of the atomic properties described in Unit 8?

It is true of the ionization energy, the electronegativity, and the covalent radius.

The ionic radius increases by 40 per cent from Li^+ to Na^+, and by 33 per cent from Na^+ to K^+, with smaller subsequent increases down the Group. Thus for an alkali metal cation Li^+ is small but Na^+ also is rather small, so there is no great disparity in Group I from the second to the third row, as there is in later Groups.

> Why do we lay so much stress on the sizes of the ions?

Because ions are the building blocks of ionic lattices, and the building blocks must fit together, as we saw in Unit 7. (We return to the effect of ionic charge when we discuss more highly charged ions.) In Unit 6 we noted some Group trends in the thermal stability of ionic solids, such as the alkali metal carbonates. The changes down the Group arise because the enthalpy of decomposition of the

21

solid depends on the lattice energy, which in turn depends on the radius of the cation for a given anion. *Try SAQ 9 on this subject.*

> SAQ 9 (Objective 6)   In each of the following cases, does the thermodynamic stability of the salts increase or decrease down the Group of the alkali metals (M)?
>
> (i)   MA relative to MB, when A^- is larger than B^-.
> (ii) M_2A relative to M_2B, when A^{2-} is larger than B^{2-}.
> (iii)   MA relative to the elements, when A^- is small.

In (iii) of SAQ 9, the trend runs counter to the trend in atomization and ionization energy down the Group (Unit 5).

Which member of Group I most readily forms a gaseous cation?

Caesium. The trend is
$$Cs > Rb > K > Na > Li.$$

In (iii) the alkali metal that most readily (in thermodynamic terms) forms a cation in a lattice with a small anion is lithium, and the trend is

$$Cs < Rb < K < Na < Li$$

This is because of the increase in ionic size down the group. Since Li^+ is small, its compound with anions that are small (and/or highly charged) tend to have high lattice energies L_0, since the opposing charges can get close together; when the anion is small, L_0 decreases steeply down the Group.

The effect of a highly charged ion is neatly shown by the nitrides. Lithium is the only alkali metal to combine directly with nitrogen gas to form a nitride, Li_3N. This is a colourless, salt-like compound. The nitride ion N^{3-} is not particularly small; it is about the size of Cl^-. But as anions go, it is very highly charged and this gives lithium nitride a high lattice energy; in this instance also L_0 decreases sharply down the Group.

Similar considerations apply to the important case of the electrochemical series of the metals in aqueous solution (Unit 5, Table 1). We discuss the order of the alkali metals in this series in Appendix 1 (White), and conclude that lithium heads this series, despite the fact it requires a large amount of energy to form a gaseous cation, because the hydration of the small Li^+ ion is very exothermic. The exothermicity of the hydration of the cation decreases down the Group as the size of the ion increases.

Is this in parallel with the lattice energy relationships discussed above?

Yes. L_0 for a given anion decreases with the increase in size of the cation.

9.6   Beryllium

Beryllium is a light metal, and rare. It is named after the alumino-silicate mineral beryl in which it is found. The gemstone emerald is beryl, coloured green by chromium.

Unlike the other metals of Groups I and II, beryllium is strong, high-melting and high-boiling, and relatively unreactive. Like the metals of Group I, and magnesium and calcium, it is extracted by electrolysis of the fused chloride. The metal has several specialized uses, e.g. in light alloys for space-craft, but it is rather expensive, and the powder is poisonous.

Beryllium resembles its Group much less than lithium does. Thus most beryllium compounds have different crystal structures from the corresponding compounds of magnesium.

Can you think of a reason for this?

Unit 6, Figure 1 shows that MgO has the NaCl lattice, as do the oxides of Ca, Sr and Ba.

The Be^{2+} ion is very small (Unit 8, Figure 32) and this leads to small coordination numbers.

22

Beryllium in BeO has CN 4, as does beryllium in crystalline BeF_2. We shall find, as we move across the second row, that these elements don't exceed a coordination number of four in their ionic and covalent compounds, with a few exceptions (e.g. the lithium halides, which all have the NaCl structure).

Mg^{2+}, Ca^{2+} etc. have CN 6 in their oxides.

Bridged chain structures in beryllium chemistry

Figure 17 shows the bridged chain structure found in $BeCl_2$, $BeBr_2$, BeH_2 and a glassy form of BeF_2. Crystalline BeF_2 has a three-dimensional structure comparable to those of ionic compounds, but when molten BeF_2 is a very poor conductor, and it often cools to a glass containing tangled chains with fluorine bridges.

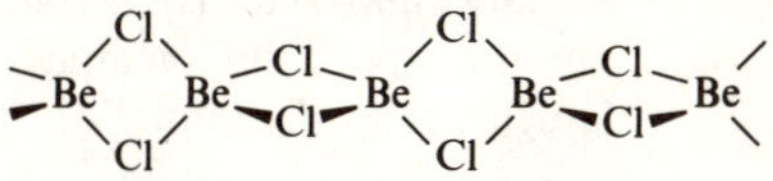

Figure 17   The chain structure of $BeCl_2$; similar chains are found in $BeBr_2$, BeH_2 and glassy BeF_2.

Notice the tetrahedral coordination of beryllium in these chains, which are covalent rather than ionic compounds.

What is the hybridization of the beryllium here?

The beryllium is sp³ hybridized.

Where have you seen chlorine bridging before?

In the structure of Al_2Cl_6 (g), shown in Unit 7, Figures 10 and 12.

Exercise 2: Making models

You can make these structures with the Home Kit balls and springs. First of all make a molecule of Al_2Cl_6(g) and this will represent also a molecule of B_2H_6 (Section 9.7.4). For the tetrahedrally coordinated atom use the balls you usually use for carbon, and for the bridging atom use the universal (14-holed) balls, since the bond angle is small. If you use these same balls for the terminal atoms, and then continue the chain as in Figure 17, you have a model of the $(BeX_2)_n$ chain.

The monomeric $BeCl_2$ molecule described in Section 9.3 is known in the gas phase at 700–800 °C. It has some historic importance, since the measurement of its vapour density showed beryllium to have an atomic mass of 9, which confirmed Mendeleev's assignment to Group II. Previously beryllium had been claimed as tervalent because of its resemblance to aluminium in chemical properties, but there was no room for such an element in Group III.

In Section 9.3 also we describe the bonding in the hypothetical molecule BeH_2, which has not been detected. It is unstable (at least at normal temperatures) with respect to the long-chain polymer $(BeH_2)_n$, just as BH_3 is unstable with respect to its dimer B_2H_6.

Figure 17 shows halogen bridging in the $BeCl_2$ macromolecule, and the hydrogen bridges in $(BeH_2)_n$ have the same shape. The bridging arises in each case because beryllium is electron deficient for the purpose of covalent bonding; it has three orbitals (one s and three p) but only two valence electrons. However, the way in which this deficiency is remedied, so that beryllium uses all four orbitals, is different in hydrogen and halogen bridging.

hydrogen and hydride bridging

In the chlorine bridge, as described in Unit 7, Section 7.4 for Al_2Cl_6, the chlorine atom uses two electron pairs to bond to two Be or Al atoms. In the hydrogen bridge, however, as in BeH_2, MgH_2 and AlH_3 (all macromolecules) or B_2H_6 (a discrete molecule), this number of electrons is not available, since hydrogen has only one valence electron. In the hydrogen bridge, one electron pair binds three atoms together, e.g. $B\diagdown^H\diagdown B$; this is called a *three-centre bond*. Of course, the other three-centre bond, $B\diagdown_H\diagdown B$, also helps to hold the borons together.*

three-centre bonds

SAQ 10 (Objective 6)   What is the significant difference, in terms of electrons, between a hydrogen bridge and a halogen bridge, as in B_2H_6 and Al_2Cl_6?

Why are the $(BeX_2)_n$ chains infinite, while B_2H_6 and Al_2Cl_6 are dimers?

* The bonding in a hydrogen bridge (or hydride bridge) is explained in terms of molecular orbital theory in Appendix 2 (Black).

All the hydride bridges are unstable with respect to reaction with water, as we shall see in the case of diborane in Section 9.7.3.

How can we understand why beryllium should form covalent bridged-chain halides, rather than ionic halides as formed by all the other metals of Groups I and II?

The Born cycle (Unit 5) shows us one way of answering this question. To form a stable ionic lattice, the metal and halogen must obtain the energy that is needed to form the ions during the exothermic steps in the reaction, of which the most important is the coming together of the cations and anions. That is to say, the expenditure on atomization and ionization is covered by the gain from lattice energy L_0 (and EA).

> How does the energy of formation of Be^{2+} compare with that for the other Group II metals? (Consult your *Data Book*.)

Therefore, we need a high lattice energy to stabilize Be^{2+} ions. L_0 is high when the internuclear distance ($r_+ + r_-$) is small, and when the cation is surrounded by a large number of anions. But Be^{2+} is too small to meet the latter condition, except with a small anion such as F^-. In crystalline BeF_2, beryllium is 4-coordinate, as we saw above. But inside a tetrahedron of chloride ions in contact, the little beryllium cation would rattle about. Its radius is only 31 pm, compared with 182 pm for Cl^-. (It is interesting that the ionic form of BeF_2 is not much more stable than the glassy form, which contains bridged chains like those of $BeCl_2$; an ionic lattice is 'only just stable', even with F^-.) Even if it were geometrically possible* to build a lattice of spherical Be^{2+} and Cl^- ions, in 1:2 ratio, and with about their usual radii, so that unlike charges are close together and like charges separated, it would be at a disadvantage in terms of energies compared with a covalent structure in which the beryllium and chlorine could get close together.

Thus the high charge density of the Be^{2+} ion (the combination of high charge and small radius) leads, paradoxically perhaps, to the formation of covalent rather than ionic bonds in beryllium compounds. The analogy with Al^{3+} is obvious.

The geometry of an ionic structure is determined by the radii and charges of the spherical ions, which must pack together as closely as possible to satisfy the electrostatic requirements.

> What are these?

In Unit 7, Section 7.4.2 we studied the principles that determine the geometry of covalent molecules.

> What is the name of this 'theory'?

These principles can be applied to some extent to giant-molecular structures, as in SAQs 12 and 16.

Thus, with ionic structures at one extreme and covalent structures at the other, there is a continuous spectrum of bond types. We found that we could visualize the transition from covalent to ionic bonding in terms of the differences in electronegativity between the bonded atoms. How can we similarly visualize the transition from ionic to covalent bonding?

Fajans' rules, which were briefly introduced in Unit 7, Section 7.4.2 and are more fully described in Unit 10, Appendix 3 (White), help us to do this. The covalent bonding in beryllium chloride forms an apt illustration of the polarization of a fairly large anion by a small and highly charged cation, as described by Fajans.

It is much higher. In round figures, ($I_1 + I_2$) is 2 700 kJ mol^{-1} for Be, as against 2 200 for Mg, falling to 1 600 for Sr. We know too that ΔH_{atm} is high for beryllium, which was described at the beginning of this Section as harder, and higher-melting and higher-boiling than the other Group II metals.

The unlike charges attract, and the like charges repel.

Valence shell (electron pair) repulsion theory.

---

* For those of you who have read about radius-ratio rules in Unit 7, Appendix 1 (Black) the minimum radius ratio r_+/r_- for 4-coordination, with all adjacent ions in contact, is 0.225. The radius ratio for Be^{2+} and F^- is just greater than this, at 0.228, but that for Be^{2+} and Cl^- is much smaller, 0.17.

There are other respects in which beryllium chemistry differs from that of the other Group II metals. Thus beryllium and its hydroxide dissolve in alkali to give the beryllate anion $Be(OH)_4^{2-}$, analogous to aluminate $Al(OH)_4^-$. In the case of beryllium metal, hydrogen gas is produced as well.

Write equations for these two reactions.

$$Be + 2H_2O + 2OH^- = Be(OH)_4^{2-} + H_2(g)$$
$$Be(OH)_2(s) + 2OH^- = Be(OH)_4^{2-}$$

On the other hand, beryllium metal, oxide and hydroxide follow the rest of Group II in dissolving in acid such as dilute HCl. They give the ion $Be(H_2O)_4^{2+}$, in which Be^{2+} is stabilized by hydration, which spreads the positive charge over a larger volume, as described at the end of Appendix 2 (White).

Beryllium is thus the only Group II metal which is amphoteric.

Clearly beryllium resembles aluminium in its chemical properties more than it resembles magnesium in its own Group. Similarly, aluminium resembles beryllium more than it resembles boron in its own Group. This is a well-known example of a *diagonal relationship* in the Periodic Table, and we shall find a corresponding relationship between boron and silicon in Unit 10.

diagonal relationship

We can understand these diagonal relationships in terms of atomic size, and ionic charge density. (Fajans' rules show that this is important, in determining bond type.) The decrease in size across the row together with an increase in ionic charge is partially compensated by the increase in size down the Group. Similarly, in Units 8 and 10, we notice that the semi-metallic elements form a diagonal band across the Periodic Table.

Although beryllium differs from its Group more than lithium does, these differences are less marked than they are for later members of the second row. The properties of beryllium form a bridge between the metal lithium, and the non-metal boron. *But before you go on to the next Section, you might like to try SAQs 11 and 12.*

SAQ 11 (Objectives 6a and 6b)   Draw up a simple table to show how beryllium chemistry differs from that of the other Group II metals; and add for comparison the corresponding information about lithium.

SAQ 12 (Objective 6)   Use the theory of valence shell electron pair repulsion (Unit 7, Section 7.4.2) to interpret or predict the geometry of the coordination of silicon by oxygen, and vice versa, in the giant-molecular compound quartz $(SiO_2)_n$. Assume that the bonding is mainly covalent, that the octet rule is obeyed, and that silicon is bound to silicon via oxygen.

Predict similarly the coordination of silicon by carbon, and vice versa, in silicon carbide SiC, known as carborundum.

9.7   Boron

Boron is not a common element, but borates such as borax are found in large deposits in several desert regions (as in California) of known volcanic activity.

Elemental boron is a giant-molecular refractory* solid, as is diamond, but more complex in structure. Remarkable cage formations, unlike anything so far discovered outside boron chemistry, are found in elemental boron, and also in boron carbide $(B_4C)_n$. The cage is a (hollow) regular icosahedron B_{12}, and these are close-packed in the structure shown in Figure 18.

refractory

The icosahedra are joined to each other by strong covalent bonds, and that is why the element is refractory; boron is nearly as hard as diamond.** Boron filaments have been developed in the USA for strengthening propeller blades, as have carbon fibres in England.

In boron carbide, the boron icosahedra are joined together by C—C—C groups.

---

* A refractory solid (like a refractory child) is one that is not easily changed. Refractories are high-melting, high-boiling, insoluble and unreactive.

** The bond energies for the B—B and C—C bonds are 301 and 347 kJ mol^{-1} respectively.

Boron carbide is also refractory, and is used as an abrasive. Other hollow polyhedra are found in borides and boranes.

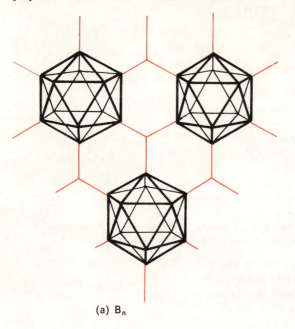

(a) B_n

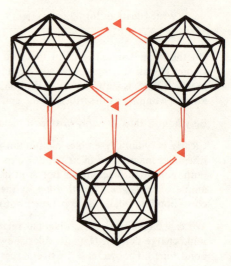

(b) $(B_4C)_n$

9.7.1 Electron deficiency in boron compounds

We call boron electron-deficient because if it forms simple electron-pair bonds with its three valence electrons, it doesn't achieve a stable octet.

In this Section and in Appendixes 2 and 3, we see various ways in which this deficiency is remedied: by acceptance of an electron pair from a Lewis base; by delocalization of π electrons from N, O or F bonded to boron; or by the formation of a 3-centre bond in the hydrogen bridge so that one electron pair holds three atoms (B—H—B) together. In the cage structures of Figures 18 and 21, some electrons are delocalized over the whole cage, as described in Appendix 3 (Black) on p. 48. In this Appendix also are described various boron-nitrogen compounds, including the giant-molecular boron nitride structures.

9.7.2 The boron halides: Lewis acids and bases

In Section 9.3, we described the bonding in BF_3. This compound illustrates several important features of boron chemistry.

First, BF_3 is a molecular compound, a gas, and this in itself proves that boron cannot form salts containing the simple cation B^{3+}. How does BF_3 prove this? B^{3+} is very small, as is shown by the plot of the ionic radii (Unit 8, Fig. 32), so it could only form a stable lattice with a very small anion (to get a high enough coordination number)*. F^- is the smallest anion. But BF_3 is a molecular compound. Therefore, there are no salts of the B^{3+} cation.

BF_3 illustrates also how the electron deficiency of boron can be made good by π-bonding. The boron atom has only three valence electrons and for maximum stability must use all four orbitals to form four bonds. In BF_3, which is flat, the fourth orbital on boron is a π orbital, and this is filled by electrons from the halogens (see SAQ 8). This stabilization by π bonding explains why the boron halides don't dimerize as BH_3 does.

This delocalization, however, is for want of something better, and BF_3 readily accepts an electron pair from a donor molecule such as ammonia NH_3 to form four σ bonds, as we show in Figure 19 from TV programme 10.

Similar $\overset{\delta-}{X_3}B—\overset{\delta+}{N}Y_3$ complexes are formed by many derivatives of borane (BH_3), including the alkyl derivatives BR_3 (such as boron trimethyl) and the halides BF_3, BCl_3, BBr_3 and BI_3, acting as *electron-pair acceptors*.

Figure 18 (a) One of the structures of elemental boron, B_n. (b) Boron carbide, $(B_4C)_n$. The small red triangles represent the C—C—C group seen end on. We give these structures because they are such remarkable designs; do not try to memorize them.

electron-pair acceptors

* If the anion is highly charged, then Fajan's rules tell us that the compound is likely to be covalent.

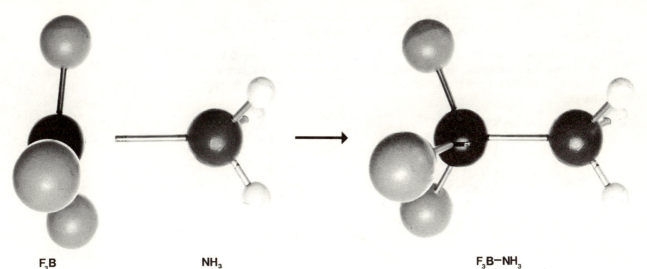

F₃B   **NH₃**   **F₃B—NH₃**

Electron-pair donors include ammonia and its derivatives such as the amines, and also water and its derivatives the ethers (ordinary ether is $(C_2H_5)_2O$).

Compounds such as $F_3B—NH_3$ and $F_3B—OH_2$ were called by G. N. Lewis *acid-base complexes*. He called BF_3 an acid because it behaves like H^+ in accepting an electron pair from NH_3, but as it is rather different from the acids we are used to, we call it a *Lewis acid*.

The electron-pair donor is called a *Lewis base,* by analogy with OH^-, which gives an electron pair to H^+. Another Lewis acid–Lewis base reaction is the formation of the fluoborate ion:

$$BF_3 + F^- \rightarrow BF_4^-$$

A fuller account of Lewis acids and bases is given in Unit 11. Ammonia-borane, $H_3\overset{\delta-}{B}—\overset{\delta+}{N}H_3$, is described in Appendix 3 (Black).

The most important Lewis base is water, and this hydrolyses all the boron halides, so that they fume in moist air.

$$BCl_3 + 3H_2O = 6HCl + B_2O_3 \text{ (and hydrated forms of this).}$$

BF_3 is, however, more resistant to hydrolysis than the other boron halides, and gives BF_4^- with the F^- formed by partial hydrolysis.

In the industrial preparation of BF_3, boric oxide is heated with a mixture of fluorspar CaF_2 (or some other fluoride mineral) and concentrated sulphuric acid, which generates HF.

Write equations for the production of HF and of BF_3.

For laboratory use, BF_3 can be obtained in cylinders, or in liquid form as the etherate $(C_2H_5)_2O—BF_3$ in which ether acts as Lewis base, but gives up boron fluoride to a stronger base.

The comparison of boron fluoride with the other halides illustrates some important principles.

The relative strengths of Lewis acids can be estimated e.g. by comparison of the enthalpies of reaction with a given base (Unit 11, Appendix 1). When this is done for the boron halides, it is always found that the Lewis acid strengths decrease in the order $BBr_3 > BCl_3 > BF_3$.

Is this the order that you would expect from the electronegativities?

But it is not, and the explanation for this unexpected inversion in order leads us again to a very important difference between the second-row elements, and the elements of later rows.

In Section 9.1.7 we showed how the second-row elements form $p\pi$ bonds much more readily than the elements of later rows. We attributed this to the rather

Figure 19 The stereochemistry of the reaction between NH₃ and BF₃, from TV programme 10.

electron-pair donors

acid-base complexes

Lewis acid

Lewis base

$CaF_2 + H_2SO_4 = CaSO_4 + 2HF$
$B_2O_3 + 6HF = 2BF_3 + 3H_2O$

No. Since fluorine is the most electronegative halogen, we might expect BF_3 to be the most avid electron acceptor.

small size of the second-row elements, allowing efficient pπ orbital overlap. In Section 9.3, we described the pπ bonding in BF_3 in detail.

Figure 20, the singularity of BF_3, shows the contrast between the short, strong B—F bond and the longer, weaker bonds of the other boron halides, due to the much greater efficiency of pπ bonding in the fluoride compared with the other halides.

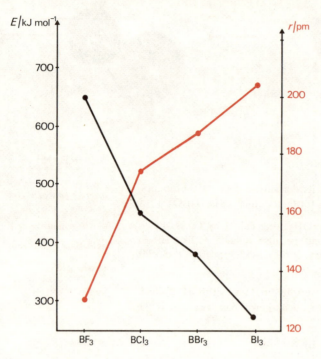

Figure 20 The singularity of BF_3. The bond energies of the boron halides, BX_3 are shown in black, and the bond lengths in red.

How then does this account for the order of Lewis acid strengths, $BBr_3 > BCl_3 > BF_3$?

In Section 9.3 and TV programme 10 we show that the π-bonding contributes significantly to the stability of BF_3, and that this contribution is sacrificed when the adduct is formed. This is because the orbital on boron that is used for π bonding in BF_3, is the one used for the B—N σ bond in the adduct F_3B—NH_3 (Figure 19). Since pπ bonding contributes very little to the stability of the heavier boron halides, they have much less to sacrifice when they form acid–base complexes. Thus, boron chloride is a better Lewis base than boron fluoride, because boron fluoride is better stabilized by its π bonding.

9.7.3 The borates

Boric acid, formerly called boracic acid, has the formula H_3BO_3. If we write this as $B(OH)_3$, we see that it is analogous to BF_3. Boric acid, like BF_3, is a Lewis acid, and in water it tends to add OH^-, so that it acts as a weak monobasic acid:

$$H_2O + B(OH)_3 \rightleftharpoons B(OH)_4^- + H^+$$

Solid boric acid has a layer structure in which the $B(OH)_3$ molecules are linked by hydrogen-bonds, shown by dotted lines

to form a network that is drawn in Evans*, p. 279. The layers are held together by van der Waals forces only, and the crystals are soft.

* R. C. Evans (1966) *An Introduction to Crystal Chemistry*, 2nd ed., Cambridge University Press.

28

There are apparently only three bonds to boron. How is the boron stabilized in this structure?

We have a similar situation here to the one in BF_3. The boron is stabilized by delocalization of two electrons from each oxygen into a system of π orbitals that covers the whole BO_3 triangle. The borates are strongly bonded and very stable.

π delocalization

In many borate structures we can see the BO_3 triangle. Many variations are possible, since oxygen may be linked to hydrogen or to another boron, or may be negatively charged.

Thus the triangles can share one corner, as in the pyroborates:

Or they can share two corners to form a ring, or chain polymers:

In the ring or chain polymers, if each negative charge is neutralized by K^+, what is the simplest formula of the borate?

The simple formula is KBO_2. The name is metaborate.

Potassium metaborate KBO_2 appears in the equation for the preparation of BF_3 (from potassium fluoborate and boric oxide) that we use in TV programme 10.

Write an equation for this reaction.

$3KBF_4 + 2B_2O_3 = 3KBO_2 + 4BF_3(g)$

The metaborate solidifies as a glass.

In Unit 10, we shall meet similar condensed oxy-anions of silicates, phosphates etc. They are called 'condensed', because the oxygen links are formed by two —OH groups coming together and splitting out water (e.g. when boric acid is heated):

oxy-anions (polyanions)

This process is reversible, many borates hydrolysing in water to give $B(OH)_3$ and $B(OH)_4^-$.

The commonest borate is borax $Na_2B_4O_7.10H_2O$, which is found in California; it is used as a mild antiseptic, and in glazing and glass-making. Pyrex is a borosilicate glass, with silicate tetrahedra (Unit 10) linked through oxygen to borate triangles and tetrahedra. Borax is a tetraborate, that is, the anion contains four borons. The structure, however, is not simple, since two of the borons are 3-coordinated and two are 4-coordinated by oxygen. If you are interested in this structure, you will find it in Evans (p. 237).

9.7.4 The boranes

The hydrides of boron are now called boranes, and the simplest one is diborane B_2H_6. Its rather unusual bonding is described in Appendix 2 (Black), and you have made a molecular model of it with the Home Kit balls and springs, in Section 9.6.

A typical preparation of diborane starts from sodium hydroborate $NaBH_4$, which is made by reduction of boric oxide with sodium hydride NaH. These simple and complex hydrides are valuable reducing agents in inorganic and organic chemistry.

$NaBH_4$ reduces BF_3 or BCl_3 in dry ether to give B_2H_6 gas, which is lost from the reaction mixture.

$$2B_2O_3 + 4NaH = NaBH_4 + 3NaBO_2$$
$$3NaBH_4 + 4BF_3 = 3NaBF_4 + 2B_2H_6(g)$$

In the early preparations of boranes, magnesium boride (formed by reduction

29

of boric oxide by magnesium at high temperatures) was treated with dilute hydrochloric acid. This gave a foul-smelling mixture of higher boranes, not including diborane itself which is too reactive under these conditions, but which could be made by thermal decomposition of tetraborane. The higher boranes are spontaneously inflammable in air, and progress in their study awaited the invention of the all-glass vacuum system by Alfred Stock in Germany in the 1920s. Stock is thus the father of borane (and silane) chemistry, and of vacuum techniques for handling gases, using refrigerants such as liquid nitrogen to move gases from one part of the apparatus to another. We use a version of Stock's techniques in TV programme 10 to show the reaction of BF_3 and NH_3, but he had to use mercury valves instead of taps, as the boranes react with tap grease.

In diborane the hydrogen bridging ensures that all four boron orbitals are used in bonding (Appendix 2); but the bridge bond is relatively weak:

$$B_2H_6(g) = 2BH_3(g) \qquad \Delta H_m^{\ominus} \approx -120 \, kJ \, mol^{-1}$$

We can compare this with the bond energies of normal (electron-pair) bonds to boron. Thus $E_m/kJ \, mol^{-1}$ is 301 for a B—B bond, or 381 for a B—H bond (cf. also 523 for B—O).

How does the bonding theory explain the weakness of the hydrogen bridge bond?

As in the boron halides, this delocalization of electrons is for want of something better. If an electron donor such as ammonia or water approaches, the bridge bond is immediately disrupted. In fact, diborane explodes in air and is immediately hydrolysed by water to form boric acid and related compounds, depending on the conditions.

The bridge bond is relatively weak, since four atoms (2B and 2H) are held together by only two electron pairs.

Perhaps you noted that the B—O bond is a particularly strong one. Why is this?

Because of pπ bonding, as in the B—F bond.

Similarly, the reaction of diborane with ammonia leads to more complex compounds than the simple ammonia-borane adduct H_3N—BH_3, and these are described in Appendix 3 (Black) on p. 48.

Higher boranes are now made from diborane, heated in the absence of air. Since World War II much money and effort has been put into their study, in the hope that they (or their alkyl derivatives) might be useful as rocket fuels. Their combustion is very exothermic, with weak bonds being broken and strong B—O and O—H bonds being formed. In the event the boranes are too expensive for this purpose, but some remarkable chemistry has been uncovered.

The structure and reactions of the boranes and their derivatives now rival organic chemistry in extent and variety. Figure 21 shows examples of the two main categories. The reactive boranes, like diborane, contain hydrogen bridges, and are air- and water-sensitive. The stable boranes include some remarkable cage structures; Figure 21 shows two hollow icosahedra, which resemble the elemental structure (Figure 18), but are discrete molecules, with a hydrogen or a substituent attached to each boron. These molecules are stabilized by the delocalization of bonding electrons in a molecular orbital that covers the whole cage in three dimensions, rather as the π bonding orbital in graphite covers a whole sheet in two dimensions.

SAQ 13 (Objective 6)   What do we mean when we say that boron is electron deficient? Or that the diborane molecule B_2H_6 is electron deficient? What is the difference in the electronic structures of BF_3 and BH_3 that is responsible for the stable forms of the fluoride and hydride being monomeric and dimeric, respectively?

SAQ 14 (Objectives 6c and 8)   Describe the important differences in geometry and the nature of the bonding, between hydrogen-bonding in water, and hydrogen bridging in diborane.

SAQ 15 (Objectives 6c and 7)   Describe some of the chemical properties of diborane that can be directly related to the three-centre bonding in the hydrogen bridge.

REACTIVE BORANES   STABLE BORANES

B_2H_6 diborane

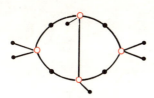

B_4H_{10} tetraborane

B_5H_9 pentaborane – 9

$B_{12}H_{12}^{2-}$ the dodecaborane ion

$B_{10}C_2H_{12}$ carborane

○ Boron   represents a hydrogen bridge

● Carbon   represents a normal electron–pair
                                   bond to hydrogen. There is one of these
• Hydrogen   attached to each vertex of the icosahedra

Figure 21 Some boranes. These structures are shown here because they are so remarkable. Do not try to memorize them (except diborane).

9.8   Carbon

With four valence electrons, and median electronegativity, carbon forms mainly covalent compounds. As described in S100, Unit 10, these may be singly or multiply bonded, simple or complex, and may contain chains, rings, or networks.

S100, Unit 8, Section 8.5.2 describes some inorganic chemistry of carbon, including its allotropy. We return to the diamond structure in Unit 10 of this Course when we compare the structures of the Group IV elements.

In Section 9.3 we described the graphite structure in terms of molecular orbital theory.

Graphite has less than two-thirds the density of diamond, so that although graphite is the stable allotrope of carbon at normal pressures, it can be turned into diamond at high pressures ($> 10^5$ atmospheres) and high temperatures (2 500 °C).

Here are some SAQs on diamond and graphite:

   SAQ 16 (Objectives 5 and 6d)   Can the coordination of carbon in graphite be justified in terms of valence shell repulsion theory (Unit 7, Section 7.4.2)?

   SAQ 17 (Objective 6d)   Diamond is a non-conductor of electricity and does not melt; it sublimes at about 3 500 °C. Graphite is commonly used for electrodes (and also sublimes without melting). Explain.

9.8.1   Carbides

Carbides are compounds in which carbon is the more electronegative partner. There are several different kinds.

Acetylides are carbides containing $^-C{\equiv}C^-$ ions, which give acetylene on

31

hydrolysis. They are formed by the alkali metals and the alkaline earth metals, and some transition metals. The best-known, perhaps, is calcium carbide, CaC_2. Calcium carbide has an interesting structure, which it shares with the peroxides of the alkaline earth metals and the superoxides of the alkaline metals (Unit 6, SAQ 25). It has a distorted NaCl lattice with the $^-C\equiv C^-$ ions all parallel to one axis, so that the unit cell is elongated in that direction.

We have already met two covalent *refractory carbides*, those of boron and silicon. Silicon carbide, SiC, is very hard and used as an abrasive, as is boron carbide. There are several forms, one of which has the zinc blende structure (Unit 7). It is, in effect, the diamond structure with every other carbon replaced by silicon.

Interstitial compounds (carbides, nitrides, borides) are described in Appendix 4 (Black) on p. 50.

9.8.2 Oxides of carbon

The common oxides of carbon are the monoxide and dioxide. We described the molecular orbitals of CO in Section 9.2. Unexpectedly, CO is a non-polar molecule; it has a very small dipole moment indeed (Unit 1, and Unit 10, Appendix 2 (White)).

Why is this lack of polarity unexpected?

We can explain these properties of CO either according to the MO or the valence bond model.

We expect the CO bond to be polar because of the electronegativity difference between carbon and oxygen, 1.0 on the Pauling scale.

Can you see how? Write down any possible Lewis structures, and consider the positions of the lone pairs.

The only Lewis structure in which carbon is tetravalent is $:\overset{-}{C}\equiv\overset{+}{O}:$

In Section 9.2 we mentioned that CO is isoelectronic with N_2. There is a lone pair on each atom (representing the electrons in the $\sigma 2s$ and $\sigma*2s$ orbitals). If you count electrons, four from the carbon atom and six from oxygen, you see that there has to be some transfer of electrons from oxygen to carbon; older formulae showed CO as $C\overset{<}{=}O$. The observed very small dipole moment shows that this transfer has taken place sufficiently to counterbalance the $\overset{+}{C}-\overset{-}{O}$ distribution that is expected on electronegativity grounds. The symmetry of the overall distribution of charges minimizes electron repulsion.

CO is one of the very few neutral oxides; it can be described as weakly acidic as it reacts with water under pressure to give formic acid, HCOOH.

In addition, CO can act as a Lewis base, forming many complexes called *carbonyls*.

$$OC\overset{\displaystyle CO}{\underset{\displaystyle CO}{-Ni\diagdown CO}}$$

Perhaps the most well-known of these is nickel carbonyl $Ni(CO)_4$ which, because it is very low-boiling (it boils at 43 °C), is used to purify nickel in the Mond process. The carbonyl is formed at 60 °C from the action of CO on the impure metal, and passed into a decomposer at 160 °C where the pure metal deposits.

SAQ 18 (Objectives 4 and 6)   What shape do you predict for the CO_2 molecule, according to valence shell electron repulsion theory? What is the bond order? The π orbitals in CO_2 resemble those in a molecule described earlier in this Unit. What shape are they?

Carbon dioxide reacts slowly with water to form a weakly acidic solution containing hydrated CO_2, and a little carbonic acid.

$$CO_2 + H_2O \rightleftharpoons CO_2 \text{ hydrate} \rightleftharpoons H_2CO_3 \rightleftharpoons HCO_3^- + H^+$$

Soda water tastes only faintly acid; this is not so much because H_2CO_3 is so weak an acid as because only a small proportion of the dissolved gas is present as carbonic acid, rather than as hydrogen-bonded hydrate. That the reaction with water is slow, is shown by the slow fading of the red colour of phenolphthalein indicator when dilute CO_2 solution is used to acidify a dilute solution of NaOH.

The atmosphere contains about 300 ppm by volume of CO_2, and this is involved in several geochemical cycles.

CO_2 is absorbed in photosynthesis; it is taken in by green plants in sunlight, and converted into carbohydrates such as sugars. These plants, if not eaten by animals or decomposed, may remain to turn into fossil fuels.

Marine organisms use chalk ($CaCO_3$) in their shells, and this deposits to form chalk beds, coral reefs, etc., and ultimately sedimentary rocks such as limestone, and metamorphic rocks such as calcite and dolomite, and marble.

The CO_2 absorbed in ground waters plays an important part in the weathering of rocks. Insoluble carbonates such as chalk are slowly taken into solution as soluble bicarbonates.

Describe the mechanism of this (cf. S100, Unit 9).

Silicate rocks (Unit 10) are very slowly decomposed, as the alkali metals are removed as soluble carbonates, and alkaline earth metals as bicarbonates, to leave mainly quartz, which is free silica, and clays containing aluminium silicates etc.

CO_2 is produced in the respiration (and this includes fermentation and decay) of animals and plants. The pH of the blood is influenced by the concentration of dissolved CO_2 (produced in the oxidation of carbohydrates etc. in the tissues), and this acts on the part of the brain that controls the lungs.

CO_2 is also produced by the combustion of fuels, etc. and the earth's power consumption is doubling every decade. The CO_2 content of the atmosphere is thought to have increased by perhaps 10 per cent in the last 150 years, and 'eco-doom' prophets have suggested that, with continued increase, the 'greenhouse' effect of CO_2 will raise the earth's surface temperature, and melt enough of the polar ice-cap to flood a significant portion of our coastal towns and cities; London, for example, is vulnerable to flooding. In reply to this it is claimed that the earth's climate undergoes many fluctuations, and there is no sign that these are being disturbed by man's efforts, which are puny in comparison with nature's. In fact, the 10-year mean temperature of the earth fell by about 0.2 °C in the 10 years to 1969, and rose by 0.4 °C in the preceding 60 years.

An 'insoluble' carbonate such as $CaCO_3$ in water is in equilibrium with a very small concentration of its ions:
$CaCO_3(s) \rightleftharpoons Ca^{2+}(aq) + CO_3^{2-}(aq)$.
If one of these is removed, e.g.
$CO_3^{2-}(aq) + H^+(aq) \rightleftharpoons HCO_3^-(aq)$,
more dissolves to take its place (Le Chatelier's principle).

9.9 Nitrogen and nitrides

Nitrogen is a versatile element. With five valence electrons, it can form nitride ions N^{3-}; or else three single covalent bonds, as in ammonia and its derivatives the amines (or four in cations such as NH_4^+). It can form $p\pi$ bonds with other second-row atoms, notably B, C, N and O; and 'odd molecules' containing an unpaired electron such as NO and NO_2. It can therefore show a variety of oxidation states, and this is explored in SAQ 20 and in the Home Experiments.

The nitrogen molecule, however, with its very high bond strength due to the triple bond (Section 9.1.4) and high ionization energy is very unreactive.* The ionization energy of molecular nitrogen is nearly as high as that of argon. Among the few reactions that take place at room temperature in which nitrogen can be 'fixed' as it is called, is combination with lithium to form a nitride Li_3N. But although *we* find the fixation of nitrogen very difficult, some bacteria, algae and yeasts can do it with ease. Leguminous plants, such as peas and beans, have in their root nodules bacteria which convert atmospheric nitrogen into nitrogenous compounds which the plants can use.

* Luckily so, for otherwise our atmospheric oxygen would disappear with some of the nitrogen to form oceans of nitric acid. The reaction of nitrogen with oxygen and water to give nitrate ion is thermodynamically favourable, but the activation energy is very high (compare the preparation of nitric acid in Section 9.9.5).

Because of the need for nitrates, as fertilizers, in explosives, and in plastics, textiles, etc., methods for the fixation of nitrogen are still being sought. Some transition metal complexes combine with nitrogen at room temperature, but there is as yet no method that can be developed industrially. The important methods all need higher temperatures; in the Haber process nitrogen and hydrogen combine at 500 °C, and in the Ostwald process nitrogen and oxygen with a catalyst give nitric oxide at 800–1 000 °C (this reaction is promoted also by lightning, and supplies some nitrate to the soil). The Haber and Ostwald processes are described below.

Many elements combine with nitrogen or ammonia on heating to form nitrides, and these are of various types. Nitrides of lithium and of the Group II metals are colourless, transparent and salt-like, and contain nitride ions N^{3-}. These nitrides are hydrolysed by water, to give the metal hydroxide and ammonia.

Boron and aluminium nitrides, BN and AlN, are no longer salt-like; they are refractories. The B—N grouping, with $3 + 5$ valence electrons, is isoelectronic with C—C, and many boron-nitrogen compounds are known that are analogous, to some extent, with carbon compounds, as described in Appendix 3 (Black). Two boron nitride structures are known and these are analogues of graphite and diamond.

Other non-metals such as carbon, silicon, sulphur and phosphorus form a variety of covalent compounds of nitrogen. Pseudo-halogens and pseudo-halides containing nitrogen are described in Unit 11, Section 11.7. They include cyanogen $(CN)_2$, cyanide ion CN^-, cyanate CNO^-, thiocyanate SCN^-, and azide N_3^-.

9.9.1 Ammonia; hydrogen-bonding

With nitrogen, we reach the first element of the second row to have a lone pair of electrons in its normal valency state. This factor makes for certain properties that are common to nitrogen, oxygen, and fluorine chemistry. Another factor these elements have in common is their electronegativity; with chlorine, they are the most electronegative elements in the Periodic Table.

In Unit 8, Section 8.8, we described the hydrogen-bonding in the hydrides as a mainly electrostatic attraction between a lone pair on the electronegative atom, and the hydrogen of a neighbouring molecule. This profoundly affects the physical properties of these compounds. We mentioned an example of this, one of the properties of ice. The ice structure is a very open one (with large holes in it) because the bonds around oxygen are directed tetrahedrally. Its density is low, and it floats on water. Few other liquids expand on freezing. In contrast to H_2O, solid H_2S has a close-packed structure, with coordination number 12.

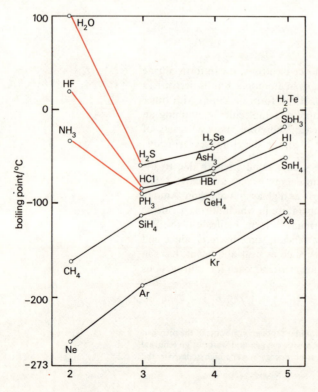

Figure 22 Boiling points of simple hydrides, showing the elevation due to hydrogen-bonding.

Figure 22 shows the effect of hydrogen-bonding on the boiling points of NH_3, H_2O, and HF.

The boiling points of the hydrides increase down the Group in Group IV (with no lone pair electrons) and in the other Groups after the first members, just as the boiling points of the noble gases increase down the Group. Other things being equal, intermolecular forces increase and boiling points rise with increasing molecular weight, as the polarizability of the molecule increases (Unit 11, Appendix 2).

In liquid NH_3, H_2O and HF, the intermolecular forces are greatly increased by hydrogen-bonding, and you can see that the boiling point of HCl is raised slightly, compared with the trends in the other groups. As we saw in Unit 8, the energy of the hydrogen-bond increases $N < O < F$. The boiling points do not increase in this sequence, perhaps because of the geometrical factor. Ammonia and water bond in three dimensions, while hydrogen fluoride forms chains.

A similar picture to Figure 22 is obtained if the enthalpies of vaporization of the hydrides at the boiling points are plotted instead of the boiling points. Thus, our oceans, lakes and rivers evaporate relatively slowly; and (as we shall see) liquid ammonia and liquid HF are useful solvents, despite their low boiling point and molecular mass.

9.9.2 Does ammonium hydroxide exist?

In TV programme 6, we considered the ammonium ion NH_4^+ as a pseudo-alkali metal cation, in between K^+ and Rb^+ in size and forming a great range of salts.

Ammonia is very soluble in water, and the solution is alkaline. Past (and present) generations of chemists were taught that the alkalinity of aqueous ammonia (by analogy with the alkalis NaOH, KOH, etc.) is due to the formation of 'ammonium hydroxide', NH_4OH, which dissociates to give NH_4^+ and OH^-. But unlike alkali metal salts, ammonium salts of strong acids, such as NH_4Cl, give an acid reaction with water. This was attributed to the formation of some undissociated 'ammonium hydroxide', considered to be a weak base.

We now have the physical methods to test such hypotheses. This is the sort of evidence they produce:

1 Ammonia forms solid hydrates of composition $NH_3 \cdot H_2O$ and $2NH_3 \cdot H_2O$ which have been studied by X-ray diffraction at low temperatures; their melting points are near the melting point of ammonia ($-78\,°C$). Like ice and solid ammonia, they are hydrogen-bonded framework structures; both contain $N-H-O$ hydrogen-bonds, and the monohydrate contains $O-H-O$ hydrogen-bonds also.

2 In aqueous ammonia, hydrogen exchanges rapidly between NH_3 and H_2O. As shown by nuclear magnetic resonance, the average $N-H-O$ hydrogen-bond has a lifetime of about 10^{-12} seconds at normal temperatures.

There is no evidence for an ammonium hydroxide molecule, and plenty of evidence for extended hydrogen-bonding. This means that we should try to explain the basic reaction of NH_3 in water, and the acid reaction of NH_4Cl in water, without the hypothetical NH_4OH.

> Write down equations to explain these reactions in terms of NH_3 and water only.

We discuss these equilibria further in Unit 11, Appendix 1, Acids and Bases. Dissociation constants are given in your *Data Book*, Section 15.

$$NH_3 + H_2O = NH_4^+(aq) + OH^-(aq)$$
$$NH_4^+ + H_2O = NH_3 + H^+ (aq)$$

You may have noticed that if a covalently bonded molecule NH_4OH existed, it would be anomalous in the bonding of the nitrogen.

> How would the bonding be anomalous?

It would require an expansion of the nitrogen octet.

9.9.3 Liquid ammonia as solvent

Liquid ammonia is a useful solvent in inorganic and organic chemistry. Although it boils as low as $-33\,°C$, the enthalpy of vaporization is high (as we mentioned above), and it is readily handled in silvered vacuum flasks (Dewar flasks).

Ammonia solvates ions well, and dissolves a variety of salts to form conducting solutions, although solubilities are different from those in water. Since ammonia is less polar than water, it dissolves many organic compounds better than water does; thus benzene and ether are soluble in liquid ammonia. Alcohols hydrogen-bond to ammonia.

Ammonia has the unique property of dissolving Group I and II metals (and one or two others, though not beryllium) reversibly, that is, the metal is recovered when the ammonia is distilled off. The solutions, however, contain the solvated cation and solvated electrons, that is,

$$Na + NH_3 \rightarrow Na^+(am) + e^-(am)*$$

Dilute solutions are blue and paramagnetic (Unit 8, Section 8.2); while concentrated solutions are metallic-looking, copper-coloured, with reduced paramagnetism, which suggests some pairing of the electrons. These solutions decompose only slowly (in the absence of catalysts) to give the amide and hydrogen gas.

$$Na + 2NH_3 = 2NaNH_2 + H_2(g)$$

These systems are however useful reducing agents. $NaNH_2$ is called sodamide; it is readily hydrolysed by water to give ammonia.

Write the equation for this hydrolysis.

$$NaNH_2 + H_2O = NaOH + NH_3$$

Pure liquid ammonia has a lower conductivity even than pure water, but it does contain a little ammonium and amide ion, as shown in Table 6.

Table 6   A comparison of the ammonia and water systems of compounds.

	Ammonia system	Water system
Self-ionization	$2NH_3 \rightleftharpoons NH_4^+(am) + NH_2^-(am)$ $K \approx 10^{-33}$ at $-33°C$	$H_2O \rightleftharpoons H^+(aq) + OH^-(aq)$ $K = 10^{-14}$ at $25°C$
acids	NH_4Cl $RCONH_2$ (organic amide)	HCl
bases	$LiNH_2$ (inorganic amide) Li_2NH (inorganic imide) Li_3N nitride	LiOH Li_2O
solvates	$Li(NH_3)_4^+$	$Li(H_2O)_4^+$
organic analogues	CH_3NH_2 (methylamine) $(CH_3)_2NH$ etc.	CH_3OH (methanol) $(CH_3)_2O$ (ether)

As we describe in Unit 11, acids on the ammonia system are compounds that give NH_4^+, and bases are compounds that give NH_2^-.

Write an equation for a neutralization reaction on the ammonia system.

$$NH_4Cl + LiNH_2 = LiCl + 2NH_3$$

One can in fact titrate a solution in liquid ammonia of NH_4Cl, as acid, against one of an alkali metal amide MNH_2, as base, with phenolphthalein as indicator. This is colourless in acid, and red in basic solution, in liquid ammonia as in water.

There are of course many organic analogues on the two systems, the ammonia system and the water system. This has led to the suggestion that planets which have ammonia in their atmosphere, but not water or oxygen, might support a

* Na^+ (am) denotes the ion Na^+ solvated by ammonia (am).

different form of life from ours. Figure 23 illustrates this possibility. The atmosphere of Jupiter, for example, consists mostly of hydrogen and helium, with a little methane and ammonia; the surface temperature approaches 30 °C and the surface pressure is over 100 atmospheres. Experiments have shown that micro-organisms including Escherichia coli (commonly found in the human intestine) can survive in conditions that simulate certain parts of the Jovian atmosphere. The ammonia men may not be far away!

Figure 23   AMMONIA! AMMONIA!

Compounds of nitrogen with the oxidation number $+3$ are usually Lewis bases, as we saw in Section 9.7.1; unless the nitrogen is bonded to an electronegative ligand such as fluorine, oxygen, or chlorine, in which case the inductive effect causes the lone-pair electrons on nitrogen to be held too strongly.

SAQ 19 (Objectives 6c and 7)   List as many Lewis acids and Lewis bases as you can think of.

9.9.4   The Haber process

As you know, ammonia is made industrially by the Haber process. This uses nitrogen obtained by the fractional distillation of liquid air. Nitrogen combines with hydrogen in the presence of a catalyst:

$$\tfrac{1}{2}N_2(g) + \tfrac{3}{2}H_2(g) \rightleftharpoons NH_3(g) \qquad \Delta H_m = -50 \text{ kJ mol}^{-1} \text{ at } 25\,°C$$

Although the equilibrium is more favourable at lower temperatures, the reaction is then too slow, so the conditions chosen must reconcile the opposing factors.

What is the effect of increasing the pressure?

Le Chatelier's principle tells us that the proportion of ammonia in the equilibrium mixture increases with increased pressure, since four molecules react to produce two. Increasing the pressure (that is, the concentration) of the reactants also speeds up the reaction (S100, Units 11 and 12).

37

Typical conditions are temperatures of 500–550 °C and pressures between 100 and 1 000 atmospheres. The gas mixture is passed through a catalyst bed of iron with some oxide.

The most important uses of ammonia are in fertilizers, and in the production of nitric acid.

9.9.5 Nitrogen oxides and oxy-acids

Nowadays, nitric acid is mostly made by the oxidation of ammonia in the Ostwald process. Special conditions are required for this, since the most thermodynamically favourable reaction, the one that takes place when ammonia burns in air, produces mainly water and nitrogen.

> Write the equation for this combustion.

$$4NH_3 \text{ (g)} + 3O_2 \text{ (g)} = 2N_2(g) + 6H_2O$$
$$(K = 10^{228} \text{ at 25 °C})$$

If, however, ammonia and air or oxygen are passed through a platinum gauze catalyst at 800 °C, nitric oxide (NO) and water are formed.

> Write the equation for this oxidation.

$$4NH_3 \text{ (g)} + 5O_2 \text{ (g)}$$
$$= 4NO \text{ (g)} + 6H_2O \text{ (g)}$$
$$K = 10^{168} \text{ at 25 °C)}$$

The oxygen used in this reaction may be a by-product of the distillation of liquid air that produces nitrogen for the Haber process.

The nitric oxide and steam are cooled and mixed with more air or oxygen which oxidizes the NO to NO_2 (nitrogen dioxide). This reacts with water to form nitric acid and NO, which is then oxidized to form NO_2, as before.

> Write the equations for these reactions.

Industrial uses of nitric acid were mentioned above.

$$2NO + O_2 = 2NO_2;$$
$$3NO_2 + H_2O = 2HNO_3 + NO$$

The nitrogen oxides, oxions and acids form a remarkable variation on the theme of σ and π bonding, which we explore in SAQ 20. Table 7 gives some distinguishing features of the more important ones.

Table 7   The more important oxides, oxions and acids of nitrogen

	b.p./ °C	Distinguishing features
N_2O	−88	Neutral, rather unreactive, stable to 500 °C
NO	−152	neutral, stable to about 1000 °C
		paramagnetic gas ; liquid and solid contain diamagnetic dimer.
		reacts with oxygen → NO_2
NO_2	21	acidic. $2NO_2 + H_2O \rightleftharpoons HNO_2 + HNO_3$
		red-brown, paramagnetic
		liquid and solid are N_2O_4, diamagnetic, colourless
N_2O_5	−32	gas is unstable → $NO_2 + O_2$
	(sublimes)	solid is ionic, $NO_2^+ NO_3^-$
		anhydride of HNO_3 ; oxidant
HNO_3	83	strong acid when dilute (aq)
		oxidant, especially when concentrated
NO_2^+		formed by conc. HNO_3 + conc. H_2SO_4
		forms salts, reacts with water → $HNO_3 + HNO_2$
		nitrates aromatic compounds etc. (i.e. introduces the nitro group, -NO_2)

SAQ 20 (Objectives 5 and 6c)   Add to the skeletal formulae the most likely arrangement of bonds and charges (i.e. draw the most likely canorical form) so that there are three bonds to each N (or four to N$^+$), and two bonds to each O (or one to O$^-$). Put in any lone pairs on nitrogen.

see skeletal formulae top of p. 39

Why is the nitrite ion bent, while the nitrogen dioxide molecule is less bent and the nitronium ion is linear?

nitrous oxide N_2O N — N — O

nitric oxide dimer N_2O_2 *dinitrogen trioxide* N_2O_3 *nitrogen dioxide dimer* N_2O_4
                                                                 (dinitrogen tetroxide)

dinitrogen pentoxide N_2O_5 *nitrosonium ion* NO^+ *nitronium* NO_2^+

                                    N — O   O — N — O

nitrogen dioxide NO_2 *nitrite* NO_2^- *hyponitrite ion* $N_2O_2^{2-}$

The dinitrogen oxides N_2O_2, N_2O_3, and N_2O_4 form an interesting series, as shown in SAQ 20. They are stable in the solid or liquid phase, and since all electrons are paired they are diamagnetic (that is, not paramagnetic). In the gas phase, the N—N bond breaks.

The order of stability is $N_2O_2 < N_2O_3 < N_2O_4$. Since NO has a bond order of $2\frac{1}{2}$, there is no gain in the number of bonds on dimerization, and so NO forms the most unstable dimer. In N_2O_4, one weak bond is gained on dimerization, and N_2O_4 molecules are found in the gas phase in equilibrium with NO_2, at room temperature.

N_2O_3 is the anhydride of nitrous acid:

acid anhydride

$$N_2O_3 + H_2O \rightleftharpoons 2HNO_2$$

N_2O_3 is an inky-blue colour, and you should catch a glimpse of it in your Home Experiments.

Nitrous oxide N_2O is not a member of the series just described, for its N—N bond is not single. Both of the structures drawn in the answer to SAQ 20 are needed to describe this molecule, for the two bond lengths are rather similar (113 pm for N—N and 119 for N—O), and the molecule has only a small electric dipole (Unit 10, Appendix 2). This may be related to the chemistry of N_2O, which is rather inert, and one of the very small group of neutral oxides.

Being non-polar, N_2O is somewhat soluble in fats. It is used nowadays as propellant in pressurized dispensers of whipped cream. A more well-known use is as an anaesthetic (nitrous oxide is laughing gas). This may have some connection with its fat solubility; other anaesthetics such as ether are fat-soluble.

9.10   Oxygen

Oxygen is the most abundant element on our planet. It forms about 50 per cent by mass of the earth's crust, much of it as silicates. The molecule is relatively inert; the bond dissociation energy is high (490 kJ mol^{-1}), and the first ionization energy (1162 kJ mol^{-1}) is similar to that of xenon, as Neil Bartlett noticed (Units 7 and 11).

Certain reactions of molecular oxygen are ubiquitous and important; respiration and combustion, for example. In the protein haemoglobin, 6-coordinate ferrous ion is held in the molecule by five nitrogens, and holds a water molecule loosely. In the lungs the water molecule is replaced by an oxygen molecule, but this is given up (and replaced by water) in body cells where the partial pressure of oxygen is low. Carbon monoxide, and cyanide ion CN^-, which are isoelectronic with nitrogen, are poisonous because they bind to the iron more strongly than oxygen does, and block its uptake.

Oxygen supports combustion, as we show in this Unit's TV programme, because the oxygen molecule has two unpaired electrons (Section 9.1.5). This makes it a di-radical. A free radical (or radical for short) is a fragment of a molecule, such as OH, and has an unpaired electron. This makes OH, for example, as reactive as a free atom of hydrogen. In the combustion process, as described in the Broadcast Notes, a flame or spark initiates the reaction by breaking some bonds to form some free atoms or radicals, and these start chain reactions in which radicals (or atoms) regenerate radicals (or atoms). Some of these reactions are branching chain reactions, in which two radicals are formed for every one consumed. The reaction gives out heat and light, and can be explosive because, as well as being exothermic, it goes very fast.

Fluorine supports combustion because the weak F—F bond breaks readily to form free atoms. The O=O bond is stronger, and breaks much less readily. Oxygen supports combustion because it is a di-radical.

9.10.1 Peroxides

The chemistry of peroxides amply illustrates the weakness of the single O—O bond. Thus organic peroxides, RO—OR, are used as initiators in polymerization reactions, as described in S100, Unit 13, Section 13.2.1. The absorption of heat or light breaks the weak O—O bond, the RO· radical adds to a monomer molecule to regenerate a radical, and the chain reaction can then continue.

Most peroxides readily decompose on heating to give the oxide and oxygen gas. We watch the decomposition of hydrogen peroxide with a catalyst in this Unit's TV programme.

9.10.2 Oxides and the Periodic Table

Except for the lighter noble gases, all the elements form oxides. These range from ionic oxides of the metals, which are usually basic (Unit 6) to the covalent oxides of the non-metals, which are usually acidic, i.e. they give acids with water. Certain oxides are covalent, but are neither acid nor alkaline; we have seen two examples in this Unit.

What are these?

Some oxides form glasses, notably B_2O_3 and SiO_2, whereas Al_2O_3 is refractory. These structures can be understood in terms of Fajans' rules (Unit 10, Appendix 3). Cations of beryllium, boron or silicon are much too small to form ionic oxides, but on the other hand boron and aluminium don't have enough valence electrons to form small covalent molecules with oxygen, and silicon doesn't form double bonds as carbon does (Unit 10). So, extended structures are formed, which are largely covalent and singly-bonded.

CO and N_2O. O_3 can perhaps be called a neutral oxide also.

As we have seen, when an element forms two or more oxides, they are usually more acidic the higher the oxidation state of the element.

Table 8   Oxides and the Periodic Table

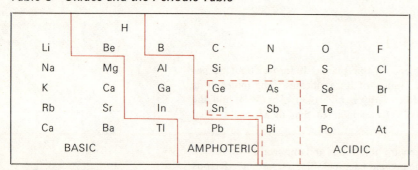

The continuous lines enclose elements with amphoteric oxides (the higher oxide, if more than one is formed).

The dotted lines enclose elements with amphoteric lower oxides.

Oxygen resembles fluorine in many of its properties, for example its electronegativity and its ability to stabilize high oxidation states in its ligands. The highest oxidation numbers of the elements are found in their oxides and fluorides, and this point is taken up again in Unit 11.

> SAQ 21 (Objective 6d) Oxygen O_2 is isoelectronic with ethylene $CH_2=CH_2$. Hydrogen peroxide H_2O_2 is isoelectronic with ethane CH_3-CH_3. Ethylene is reduced by hydrogen (with a catalyst) to form ethane. Oxygen is not reduced in this way to form hydrogen peroxide. Why not?

9.11 Summary and conclusion

We can now relate the chemistry of the second-row elements as a whole, to our earlier and later studies.

Look again at Unit 8, Figure 27. This shows, if you look for it, that the shrinkage of the covalently bound atoms across the second row is proportionately greater than that in subsequent rows of the Periodic Table. Now look at Unit 8, Figure 32, and notice just how much smaller are the ions of the second-row elements, compared with the corresponding ions of later rows. You can check with the *Data Book*, Section 13, that the radii of the second-row ions are only 2/5 to 4/5 of the radii of the corresponding third-row ions; and thereafter in the Group the radii increase much more gently.

This small size of the elements of the second row, especially the later ones in the row, lies at the root of their fundamental differences, each from the rest of its Group.

Lithium does not differ very markedly from the other alkali metals, but becomes the most electropositive metal in aqueous solution, because of the large (negative) enthalpy of hydration of its ion which outweighs the large enthalpy of formation of the gaseous cation.

Beryllium is amphoteric, in contrast to magnesium. Boron, in contrast to aluminium, is a non-metal. In the element, and in many borides and boranes, are remarkable cage structures in which boron can share its few valence electrons; these cages can only be formed by small atoms. The borates contain extended anions, comparable with the silicates.

We observed in Unit 6 the changes in the chloride of the elements, across the third row.

> Whereabouts in the third row can one see a transition from ionic to covalent bonding in the chlorides?

NaCl forms an ionic lattice, which can be described as an assemblage of charged spheres in three dimensions, packed to maximize electrostatic attraction. $MgCl_2$ and $AlCl_3$ both have layer structures in the solid, which show directed bonding that we can begin to describe as covalent, while Al_2Cl_6 vapour is molecular.

> Whereabouts in the second row from ionic to covalent chlorides does this transition come?

As we have seen (Section 9.6) beryllium chloride is covalently bonded. Similarly, we have seen that beryllium, in the second row, is amphoteric.

Beryllium and aluminium form an example of a *diagonal relationship* in the Periodic Table. Boron and the silicon provide another, as we shall see in Unit 10. The ions increase in size down the Group, but decrease in size across the row and increase in charge, and these compensations produce significant resemblances between certain diagonally related elements, as we have seen for beryllium and aluminium.

diagonal relationship

As the number of valence electrons increases across the second row, we find that multiply bonded compounds tend to be formed rather than compounds with a larger number of single bonds. Thus graphite rather than diamond is the stable form of carbon; and the carbon oxides are $:C\equiv O:$ and $O=C=O$ in contrast to silica $(SiO_2)_n$, which is a diamond analogue. The commonest nitride of carbon is cyanogen, $N\equiv C-C\equiv N$, which will be described under 'pseudohalogens' in Unit 11.

Carbon, nitrogen and oxygen form very stable π-bonded molecules. In nitrogen and oxygen chemistry these smaller molecules are preferred to the extended singly bonded systems that carbon forms in diamond, the hydrocarbons, and so on.

The formation of chains of like atoms is called *catenation*, from the Latin word catena meaning a chain, and we can now compare the elements of the second row in their ability to catenate.

catenation

How do they compare?

Carbon, of course, is the element which has this ability, *par excellence*.

Nitrogen can form chains of several like atoms in certain fluoro compounds, or if some double bonds are included, but oxygen is limited to two, with chains of three or four in certain fluoro compounds. Apart from cage formation by boron, none of the other elements of the second row can form chains of more than two like atoms. Catenation of carbon is effectively unlimited in hydrocarbon or in fluorocarbon chains, or in halocarbon chains such as polyvinyl chloride (PVC) in which a more bulky ligand such as chlorine alternates with small hydrogen or fluorine ligands*.

We have seen, in Section 9.1.7, an explanation of the failure of nitrogen and oxygen to form extended singly bonded systems by themselves. We similarly noted (also in Unit 8, Section 8.6.4) that the single bond covalent radii of oxygen, nitrogen and fluorine are larger in their homonuclear bonds, that is, in H_2N-NH_2, $HO-OH$ and $F-F$, than in their bonds to hydrogen and carbon. (These larger values are given in your *Data Book*, Sections 12 and 14; the smaller values, as found in bonds to hydrogen and carbon, are also given in your *Data Book*, Section 12.) The $N-O$, $O-F$ and $N-F$ bonds are also longer than the sums of covalent radii derived from bonds to carbon or hydrogen.

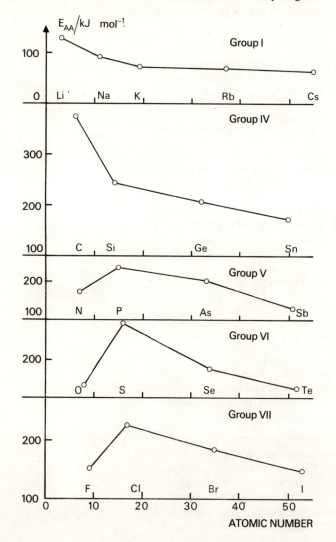

Figure 24 Homonuclear single bond energy terms.

* In Unit 10 we shall observe catenation in sulphur compounds, and to a lesser extent in compounds of silicon.

The single bond energy terms tell the same story. Figure 24 shows that the bond strengths decrease down the group in Groups I and IV (measurements are lacking in Groups II and III); while in the later Groups this pattern is followed except for the anomalously weak N—N, O—O, and F—F bonds. It is also found that the single N—O, O—F and N—F bonds are weak.

What is it that the longer, weak bonds have in common, that is not shared by the 'normal' ones C—N, C—O, C—F; N—H, O—H, H—F; P—P, S—S, Cl—Cl? How does this help to explain these observations?

The long weak bonds are between second-row atoms with lone pairs. As we saw in Section 9.1.7, because these second-row atoms are small, repulsion between adjacent long pairs squeezed close together tends to lengthen and weaken the bond. We have seen the importance of this in nitrogen and oxygen chemistry, in this Unit and its TV programme. It also plays a part in the chemistry of fluorine.

SAQ 22 (Objectives 5, 6 and 8) The 'second-row anomaly' is a name that has been given to the gross differences in chemical behaviour of the second-row elements, compared with the corresponding members of subsequent rows. List as many of these differences as you can think of, including the atomic properties.

SAQ 23 (Objective 6) Give the formulae of the elements and compounds 1–14, and classify their room temperature structures under the headings a–e. If a structure is intermediate between two of these, put down both.

a   Ionic
b   Metallic
c   Covalent-molecular
d   Covalent-giant-molecular
e   Covalent plus ionic, as in $NH_4^+Cl^-$.

1   Beryllium
2   Beryllium chloride
3   Boron
4   Lithium bromide
5   Boron bromide
6   Carborundum
7   Calcium carbide
8   Boric acid
9   Nitrogen dioxide
10   Hydrazine
11   Potassium pyroborate
12   Magnesium nitrate
13   Potassium monoxide
14   Potassium peroxide

SAQ 24 (Objectives 6b, 6c, 7 and 8) How does a Lewis acid or a Lewis base react with a hydrogen-bond? Or with a hydrogen bridge?

Appendix 1 (White)

The electrochemical series of the alkali metals

In Unit 5, Table 1, we saw that lithium heads the series of metals, in order of decreasing tendency to be oxidized by aqueous hydrogen ion:

$$M(s) + H^+(aq) = M^+(aq) + \tfrac{1}{2}H_2(g)$$

This is called the electrochemical series, since the Gibbs free energy changes in such reactions are usually obtained from measurements of electrochemical cells. This series is commonly extended to include other redox reactions in aqueous solution, that is, reactions in which reagents other than metals reduce $H^+(aq)$, or are reduced by $H_2(g)$, or in which hydrogen is not involved. The series is a useful predictor of oxidation-reduction reactions in water.

As we saw in Unit 5, the order of the alkali metals in this series is different from their order for reaction with other oxidizing agents, such as oxygen (Table 3 on p. 12) or chlorine (Table 10 on p. 28). How does this order in aqueous solution arise? Unit 5, Section 7 shows how such a question can be answered in thermodynamic terms.

> How is this, in principle?

We can draw a thermochemical cycle, which uses Hess' law to break down the reaction into a number of imagined steps (such as the loss of an electron from a gaseous atom) for which the enthalpy change is readily identified (in this case it is I_1 the first ionization energy). Some of these enthalpy changes are then properties of the metal alone (I_1 for example) and we can then examine how they vary in the Group of the alkali metals.

Figure 25 shows the thermodynamic cycle for lithium in aqueous acid (under standard conditions) with the other alkali metals for comparison. All enthalpy changes referred to here are molar quantities, and the subscript m should be understood. This is how to find your way around Figure 25:

1 When the system has to take in energy, we go up the diagram.

2 When the system gives out energy, we go down the diagram.

3 We *start* with solid metal M and aqueous acid in the bottom left-hand corner.

4 We *end* with the metal in solution (as $M^+(aq)$), and hydrogen gas, in the bottom right-hand corner.

5 The red arrow, connecting the starting and finishing levels directly, represents the overall heat of formation of the aqueous metal cation, $\Delta H_f^\ominus(M_{aq}^+)$.

6 Now we break down this overall process into simple steps, as in a Born-Haber cycle. Let us concentrate on lithium and ignore the other alkali metals shown in red.

7 To go from solid metal to gaseous metal requires energy. We go up (by $\Delta H_{atm}^\ominus$).

8 To go from gaseous metal to gaseous metal ion . . . up or down?
> . . . by how much?

Figure 25 shows that we go up, the enthalpy increasing by I_1.

9 Now we can see how the other alkali metals fare in comparison.

> Which alkali metal needs the largest amount of energy in order to form a gaseous cation?

This is lithium, which forms the smallest cation (Unit 8, Section 8.7); the ease of formation of the gaseous cation is Cs > Rb > K > Na > Li (Unit 5, Section 5.7).

10 If now the gaseous metal ion is 'put in water', so that its positive charge attracts the negative ends of water dipoles, energy is given out. $\Delta H_{hyd}^\ominus$ is the enthalpy of formation of an aqueous ion. (The way in which the total enthalpy change is apportioned between cation and anion is described in your *Data Book*,

44

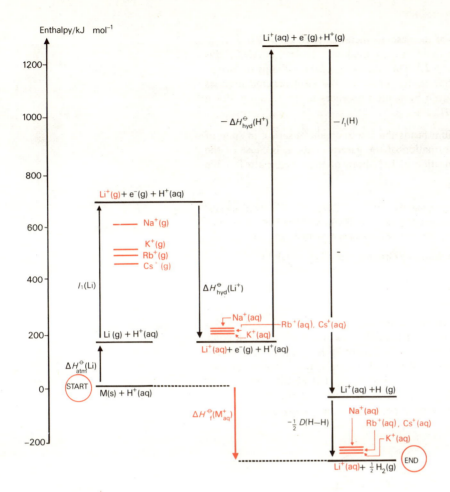

Enthalpy/kJ mol^{-1}

Figure 25 *Electrochemical series of the alkali metals in aqueous solution. The values of $\Delta H^{\ominus}$ are compared on a scale of energy, on which the standard enthalpies of formation $\Delta H_f^{\ominus}$ of M(s) and of H^+ (aq) are both zero. By this means, the relative positions of the standard enthalpies $H^{\ominus}$ for M^+ (aq) + $\frac{1}{2}H_2(g)$ in the diagram give us the values of $\Delta H^{\ominus}$ for the overall reaction. Similarly the relative positions of the standard enthalpies for M^+ (g) + H^+ (aq) on the diagram give the relative values of $\Delta H^{\ominus}$ for the formation of gaseous metal cations.*

Section 7.) $\Delta H_{hyd}^{\ominus}$ is thus analogous to the energy L_0 that is given out when a cation forms a lattice with anions (Units 5 and 6).

Which alkali metal gives out the most energy in the hydration of its cation?

Again it is lithium, the smallest, since the opposing charges can get closest together; and the exothermicity decreases down the Group.

11 Figure 25 shows, however, that the rate of variation down the Group of $\Delta H_{hyd}^{\ominus}$ is different from the variation of I_1, for in this hydration step the alkali metals are jumbled up and lose their Group order. The new order is the resultant of rather small differences, down the Group, between the rather large quantities $(\Delta H_{atm}^{\ominus} + I_1)$ on the one hand, and $\Delta H_{hyd}^{\ominus}$ on the other.

But all the same, lithium forms the most stable aqueous cation, of the alkali metals, the order being Li > K > Rb = Cs > Na.

These are all the steps that depend on the metal. In the three remaining steps the oxidant H$^+$(aq) forms a gaseous ion, which receives the electron from the metal to form a gaseous atom, which then forms half a gaseous molecule.

Write down the enthalpy changes, with sign, in these steps, before checking with the right-hand side of Figure 25.

The steps are:

12 A large endothermic step to form the gaseous ion H$^+$.

13 An even larger exothermic step to form the gaseous atom H.

14 An exothermic step to form (half of the) gaseous molecule H$_2$. Since the steps 11–13 are independent of the metal, the order of the alkali metals at the end is the same as the one (step (11) above) in the middle of the diagram, namely Li > K > Rb = Cs > Na, the order of the electrochemical series.

Now summarize the main points of our argument about the order of the alkali metals in the electrochemical series, and compare your summary with the one given below.

Summary

The order of ease of formation of the gaseous metal cations, Cs > Rb > K > Na > Li, is upset by the group trend in the enthalpy of hydration $\Delta H_{hyd}^{\ominus}$, this being of opposite sign to $(\Delta H_{atm}^{\ominus} + I_1)$. Thus the order of the enthalpy of formation of the aqueous cation, that is, the order of the electrochemical series Li > K > Rb = Cs > Na, is given by small differences in the variation down the Group of $\Delta H_{hyd}^{\ominus}$, and of $(\Delta H_{atm}^{\ominus} + I_1)$.

We can say, however, that lithium heads the electrochemical series, in spite of its relatively high enthalpy of formation of the gaseous cation, because of the large (negative) enthalpy of hydration of Li^+, both of these being attributable to the small size of lithium.

SAQ 25 (Objective 6a)   Determine the values of $\Delta H_{hyd}^{\ominus}$ for the alkali metals, using information given above and in your *Data Book*, Sections 7 and 9. A value of $\Delta H_{hyd}^{\ominus} = -1\,090$ kJ mol^{-1} has been estimated for the hydrogen ion.

How do these values change down the Group of the alkali metals? Explain.

Hydration

Hydration is an example of solvation, the binding of solvent to solute molecules or ions. This may be simple electrostatic attraction, as between ions and a polar solvent, or weak covalent bonding, or both. Figure 26 is an illustration of the electrostatic model as it applies to the alkali metal cations drawn in two dimensions. The water molecule is not a simple dipole, but we can imagine it as one. The negative ends of the dipoles are attracted towards the cation, the number of nearest neighbours of the cation depending on the relative sizes of cation and solvent molecule.

hydration

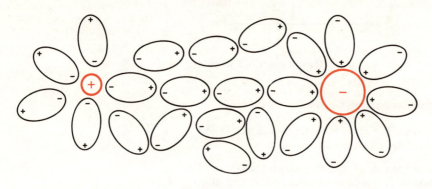

Figure 26   The solvation of a cation and an anion.

Where have you met this kind of geometry?

In Unit 6, Section 6.2.6, we observed the hydration of Mg^{2+} in crystalline Epsom salts and, in Unit 7, we related the coordination number of a cation to the relative sizes of cation and anion.

Li^+ in water has about four solvent molecules as nearest neighbours, and further molecules cluster loosely about them, with hydrogen bonding. If you imagine adjacent positive and negative charges as partially neutralizing each other, then you can see that the ionic charge is effectively spread out over the solvent cluster. This partially dissociates and re-forms, with the motion of the solvent molecules, depending on temperature.

As to the energy of the interaction, there is again a parallel with solid lattices (as described in Unit 7, Section 7.2). We expect, as we find, $\Delta H_{hyd}^{\ominus}$ to become smaller (less negative) as the cationic radius increases; this is also true of the lattice energy for a given anion.

Appendix 2 (Black)

Boron hydrides

The simplest boron hydride observed is diborane B_2H_6. Its structure has been determined spectroscopically. The molecule has four hydrogen atoms and two boron atoms in a plane with two hydrogen atoms above and below the plane (Figure 27).

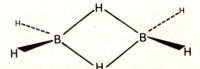

Figure 27 *The structure of diborane.*

How many electron pairs are needed for this structure?

There is something else peculiar about this structure. The hydrogen atoms form bridges between the boron atoms. The bridging hydrogen atoms have a valency of 2. You have already met bridge structures like this in Unit 7; the dimer Al_2Cl_6 has essentially the same geometry. The structure Al_2Cl_6 was explained in Unit 7 with dative bonds, a possibility not open to diborane. Apparently diborane has insufficient electrons to exist. We call compounds like this *electron deficient*:

If we allow two pairs per bond, we need 8 pairs. However, only 12 valence electrons are available.

From the structure, it appears that the terminal B—H bond is a normal σ-bond, and the bond angles suggest some sort of sp hybridization of the boron orbitals. There are two —BH_2 groups joined by hydrogen bridges. Each boron has one electron to contribute to the bridging bonds.

If the boron atoms are sp^3 hybridized, two of the hybrid orbitals are directed inwards and overlap the 1s orbitals of the hydrogen atoms. If we now combine the three atoms B—H—B, we get a new set of molecular orbitals (Fig. 28).

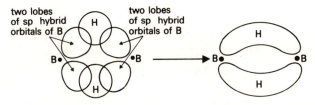

Figure 28 *The formation of two three-centre bonding orbitals.*

Three orbitals now overlap. Not surprisingly, the three atomic orbitals produce three molecular orbitals: one bonding; one antibonding; and one essentially non-bonding, i.e. making little or no contribution to the molecule's stability. These orbitals, unlike any we have considered so far, are three-centre orbitals. From Figure 28 it is apparent that there are two sets of these orbitals. So we have two bonding, two non-bonding and two anti-bonding orbitals. With four electrons available, the two lowest energy orbitals, the bonding orbitals, are filled.

three-centre orbitals

Three-centre bonds of this type are not common, but hydrogen bridges are found in the hydrides of beryllium (Section 9.6), boron (Section 9.7), magnesium and aluminium. Some of the more complex boron hydrides involve three-centre bonds between three boron atoms, as described in Appendix 3. There is now no reason to describe these structures as electron deficient in molecular orbital terms. All the bonding orbitals are occupied, and the theory accounts quite naturally for a type of structure which is otherwise very difficult to explain.

three-centre bonds

Appendix 3 (Black)

Boranes and boron-nitrogen compounds

The structure of the higher boranes is quite unusual. In many of them, a three centre bond is formed by a regular triangle of three boron atoms. Three hybrid atomic orbitals (one from each boron) overlap in the centre of the triangle to form a molecular orbital as in Figure 29. This contains an electron pair which holds the three boron atoms together.

Triangles of boron atoms are present in many boron structures. You have seen some already: the icosahedra of Figure 18, which are found in elemental boron and in boron carbide, have triangular faces. Because each edge of the figure is shared by two faces, the delocalized electrons are free to move from one face to the next, and then to the next, and so on. The result is a molecular orbital covering the whole icosahedron. This orbital is near spherical, just as the π bonding orbital in benzene is near circular, and just as the π bonding orbitals in the graphite structure are infinite sheets (Section 9.3). This delocalized bonding adds greatly to the stability of these structures.

In the boranes (Fig. 21), boron is bound to at least one hydrogen by a normal electron-pair bond, and to other borons by hydrogen bridges, or by multi-centre boron bonds. Thus, the boranes fall into two types. The compounds with hydrogen bridges are called reactive boranes, because like diborane they are sensitive to air and moisture. Some of these are icosahedral fragments, with hydrogen-bonds at most of the edges.

On the other hand, the cage boranes, with no hydrogen bridges, are remarkably stable. The salt of the icosahedral anion, $2K^+ \ B_{12}H_{12}^{2-}$, can be heated to 600 °C without decomposition and can be recrystallized from water, which would have greatly amazed a boron hydride chemist of 20 years ago.

A great range of polyhedra has been made, and many derivatives. The hydrogen attached to the cage can be substituted in the same way as aromatic hydrogen in organic chemistry; many cage borane halides have been made, for example B_4Cl_4, B_8Cl_8, and $B_{12}Cl_{12}$, in which the boron cages are, respectively, a tetrahedron, a square anti-prism, and an icosahedron. (If you take a cube, and turn one face through 45° relative to the opposite face, the result is a square anti-prism.)

Other atoms can help to form part of the boron cage; Figure 21 shows a carborane, with two carbons in the boron framework.

There is an extensive chemistry, also, of boron in combination with nitrogen.

The simplest amine-borane is H_3N-BH_3. This compound is not obtained from diborane and ammonia, which react to give more complex products, but can be made from ammonium chloride and lithium hydroborate in ether solution:

$$NH_4Cl + LiBH_4 = LiCl + H_3N-BH_3 + H_2$$

Ammonia-borane is isoelectronic with a hydrocarbon.

Which one?

The combination of boron, with three valence electrons, and nitrogen, with five, is isoelectronic with a pair of carbons with four each, and there are many compounds that correspond in this way.

Ammonia-borane is a solid at room temperature and only slightly volatile, while ethane is a gas.

Can you think why H_3N-BH_3 molecules should be more cohesive than are CH_3-CH_3 molecules?

Since nitrogen contributes two more electrons to the bonding than boron does, the bond is polar, $\overset{-}{B}-\overset{+}{N}$. (As nitrogen is more electronegative than boron, the charges are fractional, $\delta+$ and $\delta-$, but they are often written simply as + and

Figure 29 Delocalized molecular orbital in the triangular face of a borane cage.

Ethane, CH_3-CH_3.

—.) This dipole in the molecule greatly increases the intermolecular forces, so that ammonia-borane is a solid at room temperature, and $H_3C\!-\!CH_3$ is a gas.

If diborane and ammonia are heated together, they produce borazine (Fig. 30), which is more conveniently prepared from NH_4Cl and $LiBH_4$ heated together. Boron nitride can be made by heating a mixture of B_2O_3 (boric oxide) and NH_4Cl, or by strongly heating any one of a range of boron-nitrogen compounds, to eliminate hydrogen, water, or HF, etc.

Borazine has been called 'inorganic benzene', for it is flat, and has a six-centre π bond. The bonding is analogous to the bonding in BF_3 and the borates, in that an electron pair on nitrogen is delocalized into the fourth boron orbital. But again, this is for want of something better, and borazine is readily hydrolysed to boric acid and ammonia, with some hydrogen.

Write an equation for this hydrolysis.

$$(BHNH)_3 + 9H_2O = 3H_3BO_3 + 3NH_3 + 3H_2$$

Hexagonal boron nitride is a refractory solid with a layer structure; it is slippery, though not as good a lubricant as graphite; it conducts electricity, but less well than graphite. There is a significant difference in their structures (Fig. 30). In graphite, the horizontal layers are staggered, and half the carbons lie above ring centres in the next sheet; while in boron nitride, each boron lies directly above nitrogen in the next sheet, and vice versa.

This brings the opposite charges in different layers as close together as possible. Thus, there is a little ionic bonding between the layers, which explains the lower lubricity compared with graphite. The metallic character is reduced, since the electrons are more localized.

$\overset{-}{H_3B}\!-\!\overset{+}{NH_3}$ ammonia borane	$H_3C\!-\!CH_3$ ethane
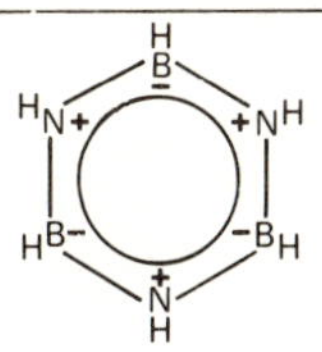 borazine	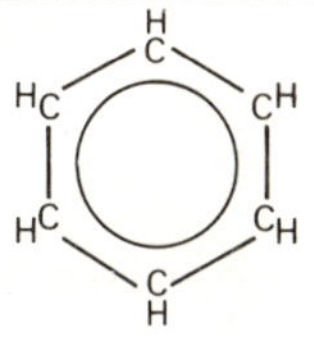 benzene
hexagonal boron nitride $(BN)_n$ (white graphite)	graphite
borazon $(BN)_n$	diamond

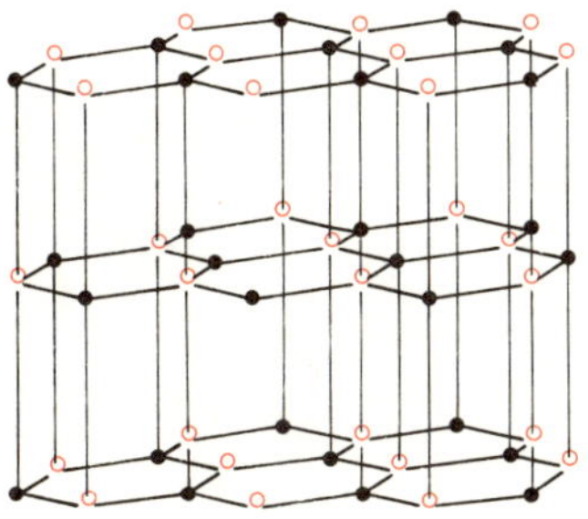

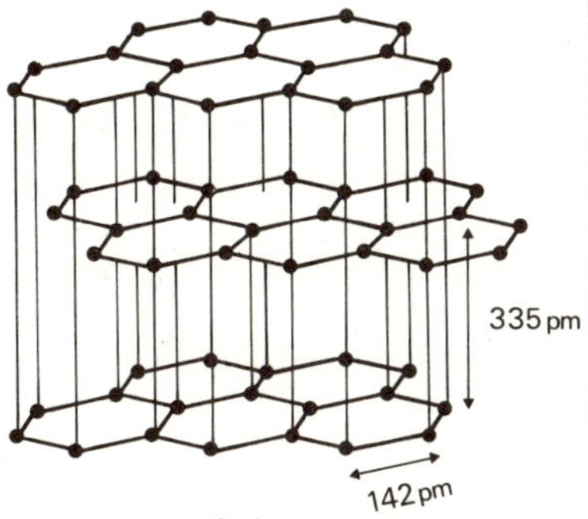

335 pm

142 pm

• Boron
○ Nitrogen

• Carbon

Figure 30 Boron-nitrogen compounds and their carbon analogues.

If hexagonal boron nitride is heated to a high temperature with an appropriate catalyst, it changes into a diamond-like form, called borazon. This is about as expensive to make as synthetic diamonds, but since it is comparable to diamond in hardness and more stable at high temperatures, it may be of use in industry.

Appendix 4 (Black)

Interstitial compounds

In Unit 8, Section 8.8.1, we mentioned the interstitial hydrides formed by the transition metals, in which the hydrogens occupy tetrahedral or octahedral holes in the close-packed metal lattice.

Figure 31 shows the clusters of spheres in close-packing; if you compare them with Unit 3, Figure 4 you will see how they belong in two close-packed layers, one red and one black.

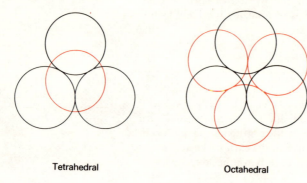

Tetrahedral   Octahedral

Figure 31   A tetrahedral cluster and an octahedral cluster in the close-packing of spheres.

Exercise 3: Models

Make each of these clusters with the polystyrene balls and glue in your Home Experiment Kit (you may wish to do this by taking apart a previous model).

You will see that the first cluster is a regular tetrahedron, and the second a regular octahedron. Now we need to compare the sizes of the holes in the middle.

In Unit 7, Appendix 1 (Black), we compared the limiting radius ratios for octahedral and tetrahedral coordination.

Write in from Unit 7 the limiting radius ratio r_I/r_M for octahedral coordination, where r_M is the radius of the metal atom, and r_I the radius of the interstitial atom.

For the octahedral cluster, r_I/r_M lies in the range 0.414–0.73.

octahedral hole

For the tetrahedral cluster, r_I/r_M lies in the range 0.225–0.414 (Evans, p. 43).

tetrahedral hole

The metals in the first transition series have atomic radii (in the metallic state) of 125–150 pm, so that for tetrahedral holes r_I can be less than 30 pm, and for octahedral holes r_I can be as much as 110 pm, as a rough guide.

As a rough guide to the sizes of interstitial atoms, we can look at the covalent radii r_c in your *Data Book*, Section 12. The values are 37 pm for hydrogen, 80 for boron, 77 for carbon, and 74 for nitrogen. With these figures, we can understand how second-row elements can form interstitial compounds as hydrogen does; but whereas hydrogen occupies tetrahedral or octahedral holes, the second-row atoms occupy octahedral holes.

Many transition metals form *interstitial carbides*, which look metallic, and conduct electricity, but are high-melting and hard. These carbides are akin to alloys, and the carbon may have some cationic character; that is, it may shed some valence electrons (as the metal does) into the delocalized orbitals which hold the 'electron sea' which gives the metal many of its characteristic properties (Unit 10). In some carbides, carbon may accept electrons from the metal.

But as you know, atoms get smaller across the row of the Periodic Table. From chromium onwards in the first transition series, the metal atoms are too small for the carbons to occupy octahedral holes in the close-packed lattice, and this is distorted in the carbide. Such carbides have properties intermediate between those of the ionic carbides and the typically interstitial carbides.

The effect on the physical properties of a metal of the incorporation of a small proportion of interstitial atoms is of great industrial importance, in the manufacture of steels, for example.

Boron and nitrogen also form interstitial compounds, in which the metal lattice may be expanded to accommodate the second-row atom. Like the corresponding carbides, the borides and nitrides are high melting, hard, and inert. They conduct electricity, and are not necessarily stoichiometric, and are thus comparable to alloys.

SAQ answers and comments

SAQ 1   For N_2^+, the energy sequence of orbitals is that shown in Figure 5, the same as for N_2. The ion has nine valence electrons so the electronic configuration is $(\sigma 2s)^2$ $(\sigma^*2s)^2$ $(\pi 2p)^4$ $(\sigma 2p)^1$. With a net excess of five electrons in bonding orbitals, the bond order is $2\frac{1}{2}$.

The O_2^- ion has 13 valence electrons and the relevant energy sequence is shown in Figure 7 (the same as O_2). The electronic configuration is therefore $(\sigma 2s)^2$ $(\sigma^*2s)^2$ $(\pi 2p)^4$ $(\sigma 2p)^2$ $(\pi^*2p)^3$, giving an excess of three bonding electrons and a bond order of $1\frac{1}{2}$. It is interesting that in the case of N_2^+, spectroscopic evidence confirms that the highest occupied orbital is the $\sigma 2p$. Spectroscopy also supports our conclusions about O_2^-; the bond length increases from 112 pm in O_2 to 126 pm in O_2^-.

SAQ 2 Ne_2 has 16 valence electrons, so that the 8 molecular orbitals in Figure 3 are fully occupied. The bond order is therefore zero and we should not expect the molecule to be more stable than the two atoms. In fact, Ne_2 has never been observed. According to more precise reasoning than the simplified approach we have adopted, it is found that the destabilizing effect of an antibonding orbital (e.g. π^*2p) on the molecule is greater than the stabilizing effect of the corresponding bonding orbital (e.g. $\pi 2p$). So the molecule is predicted to be *less* stable than the individual atoms.

SAQ 3   With an excess of five bonding electrons, NO has a bond order of $2\frac{1}{2}$. NO^+ has a total of ten valence electrons, one less than NO, which means that the electron in the highest orbital (σ^*2p) in the relevant energy diagram (Fig. 8) is removed. Removal of an electron from an antibonding orbital *increases* the bond order from $2\frac{1}{2}$ to 3. The electronic configuration of NO^+ is $(\sigma 2s)^2$ $(\sigma^*2s)^2$ $(\pi 2p)^4$ $(\sigma 2p)^2$. We therefore expect the bond in NO^+ to be shorter and stronger than in NO.

It is interesting that there is no general rule relating changes in bond order to loss or gain of an electron. The resultant bond order in the ion depends on the orbital from which the ion is removed. Contrast the formation of the two ions NO^+ and N_2^+ (SAQ 1).

SAQ 4   Atomic oxygen can be written with the valence electronic configuration $2s^2$ $2p_x$ $2p_y$ $2p_z^2$, since all three 2p orbitals are degenerate.

If the half-filled 2p orbitals are used to form molecular orbitals (σ bonds) between adjacent oxygen atoms, a bond angle of $90°$ is expected (compare this with the predicted bond angle for H_2O in Unit 8, Fig. 21). A better estimate of the bond angle is obtained if the central oxygen atom is sp^2 hybridized ($120°$ is then predicted). Two of the three sp^2 hybrid orbitals overlap with the 2p orbitals of the terminal oxygen atoms to produce σ bonds. The third sp^2 hybrid orbital contains the lone pair of electrons (Fig. 32).

The remaining 2p orbital (fully occupied $2p_z$) on the central oxygen atom is perpendicular to the plane of the molecule and overlaps with the parallel half-filled $2p_z$ orbitals on the terminal atoms. Combination of these three atomic orbitals produces three molecular orbitals: one bonding, one non-bonding and one anti-bonding. The four available electrons occupy the first two of these. The bonding orbital is therefore a π orbital spread over the three oxygen atoms in a similar way to the π bonding orbital in BF_3 (Fig. 12). Ozone is therefore predicted to have a bond order of $1\frac{1}{2}$. In molecular orbital theory, fractional bond orders emerge quite naturally and the difficult concept of resonance proves to be unnecessary.

SAQ 5   sp hybridization of the central atom produces two linear orbitals which overlap with one of the sp^2 hybrid orbitals of each of the terminal carbon atoms, to give σ bonds. The three carbon atoms therefore lie in a straight line and the $\overset{\frown}{C-C-C}$ angle is $180°$. If you are uncertain why this angle is $180°$, look at the beginning of Section 8.3 which describes sp hybridization in BeH_2.

sp^2 hybridization of the terminal carbon atoms leads us to expect the $\overset{\frown}{H-C-H}$ angle to be $120°$.

SAQ 6 sp^2 hybridization of the terminal carbon atoms leaves each of them with a 2p orbital not used in formation of sp^2 orbitals. These 2p orbitals are each occupied by one electron and are perpendicular to the trigonal sp^2 plane of the CH_2 groups.

The central carbon atom which is sp hybridized has two 2p orbitals ($2p_y$ and $2p_z$) not used in hybridization. Each contains one electron. These two orbitals overlap with the unhybridized orbitals of the terminal carbon atoms to produce π orbitals. Because these

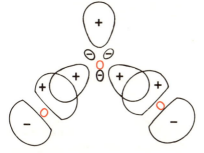

Figure 32   Overlapping orbitals in ozone.

2p orbitals of the central atom are perpendicular to each other, the 2p orbitals of the terminal atoms must also be perpendicular to each other for overlap to occur. Figure 33 shows how this comes about. The overlapping atomic orbitals are shown on the left of the Figure; the $2p_z$ of the central atom is shown overlapping with the 2p orbital of the left-hand atom at the top, and the $2p_y$ of the central atom overlaps with the 2p of the right-hand atom at the bottom. On the right of the Figure the two molecular orbitals are shown.

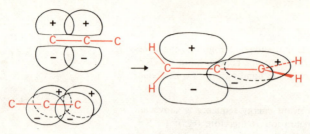

Figure 33   The formation of π bonding orbitals in allene.

To produce the overlap necessary in Figure 33 the two CH_2 groups must be perpendicular to each other as shown.

SAQ 7   The bond angles in benzene are readily accounted for by sp^2 hybrid orbitals which are involved in σ bond formation. Additionally a system of π bonds is formed by the $2p_z$ orbitals, which are perpendicular to the sp^2 plane. Read Evans, Section 4.28 on Benzene, pp. 77–8, for commentary on the bonding in benzene.

SAQ 8   In the molecular orbital formulation, the halogen is sp^2 hybridized (as boron is). Of the three sp^2 orbitals, one is used in a σ bond to boron, and two hold lone pairs. The fourth orbital contributes to the π orbital. In the resonance formulation:

i.e. a double bond resonates round the three positions. Thus the bond order is $1\frac{1}{3}$ (i.e., a single bond plus one-third of a double bond). This is matched by the MO formulation, since we mentioned in Section 9.3 that (of the 6 π electrons), one pair is bonding, and two pairs are non-bonding. One bonding pair for three bonds is $\frac{1}{3}$ of a π bond each plus the σ bond.

SAQ 9
(i)   It increases. Unit 6, Section 6.3 gives the example in which A is the tri-iodide ion I_3^- and B is I^-.
(ii)   It increases. Unit 6, Section 6.3 gives the examples of the carbonate or the peroxide relative to the simple oxide.
(iii)   It decreases. Unit 6, Section 6.3 describes the case of the simple hydrides MA.

SAQ 10   The root of the difference between hydrogen and halogen bridging is that hydrogen has only one valence electron, but a halogen has seven.

The hydrogen bridge B⟨ᴴ⟩B is held together by one electron pair only, in a three-centre bond. In the halogen bridge Al⟨ᶜˡ⟩Al, the three atoms are held together by two electron pairs; there are two bonds between the three atoms.

The $(BeX_2)_n$ molecules are infinite because beryllium lacks the valence electron that boron and aluminium use in B_2H_6 and Al_2Cl_6 to form the normal electron-pair bonds with the terminal atoms.

SAQ 11   Your table may look something like this:
√   indicates a resemblance (broadly speaking)* to the rest of the Group, lithium to Group I, beryllium to Group II.

×   indicates a difference (broadly speaking)* from the rest of the Group.

	Rare metal	Widely distributed	Made by electrolysis of chloride	High m.p. and b.p.	CN in oxide	Chloride	Hydride	Amphoteric
Li	Yes ×	Yes √	Yes √	No √	6 √	Ionic √	Ionic √	No √
Be	Yes ×	No ×	Yes √	Yes ×	4 ×	Covalent ×	Bridged ×	Yes ×

* Rb and Cs are very rare, and Cs is not widely distributed.
Mg forms a bridged hydride.

Tables of this sort are often useful in descriptive inorganic chemistry. Note that there are nearly all ticks √ for lithium, nearly all crosses × for beryllium.

SAQ 12   Silicon has four valence electrons, and completes its octet in quartz, SiO_2, by forming four bonds to oxygen. With four electron pairs in the valence shell of silicon, the bonds are directed tetrahedrally.

Oxygen has six valence electrons, and completes its octet by forming two bonds to silicon. Thus the coordination is 4:2.

Of the four electron pairs in the valence shell of oxygen, two are bonding pairs and two are lone pairs, so the bonds might be expected to be tetrahedrally disposed, i.e. the bond angle Si-O-Si to be 109.5° (or even slightly less, since in isolated molecules, lone pairs take up more room in the valence shell than bonding pairs do; see Unit 7, Section 7.4.2). In the commonest form of quartz, however, the bond angle at oxygen is 142°.

Can you think of an explanation for this size of bond angle at oxygen?

In carborundum, SiC, each atom has four valence electrons and forms four covalent bonds. They are directed tetrahedrally, with coordination 4:4. Silicon carbide crystallizes in various forms, resembling the ZnS structures, of which the zinc-blende structure was described in Unit 7, Section 7.1.2.

The electronegativity difference between silicon and oxygen is 1.7, and according to Evans (p. 157) this corresponds to about 40 per cent ionic character. In an ionic structure we might expect to find oxygen on the line joining silicon atoms, i.e. a bond angle of 180°. The observed angle is thus a compromise between the ionic and the covalent.

SAQ 13   Boron is electron deficient in that if it forms electron-pair bonds using its three valence electrons, it is still one pair short of a stable octet; one boron orbital remains empty. B_2H_6 is electron deficient in having only 12 electrons in the valence shell (3 from each boron, 1 from each hydrogen) and 7 electron pairs would be needed if all the bonds were normal single bonds, as in ethane C_2H_6.

In BF_3, boron uses three sp^2 orbitals to form σ bonds to fluorine, and its fourth orbital is a π orbital which contains electrons, delocalized from fluorine, which would otherwise be lone-pair electrons. This possibility is not available to BH_3, as hydrogen only has one electron. But if BH_3 dimerizes, the four boron sp^3 orbitals can be used for bonding; two for 3-centre bonds forming the hydrogen bridges, and two for normal electron-pair bonds to the terminal hydrogens.

Fluoride bridges are known, as we saw in the case of glassy beryllium fluoride (Section 9.6). Presumably it is easier for boron, with three valence electrons, to fill its valence shell by π bonding, than it is for beryllium with only two.

SAQ 14   Some of the information needed for this answer is in Unit 8, Section 8.8.1.

In hydrogen bonding, as in water, the hydrogen lies on the line of centres of the bonded oxygen atoms. It is nearer to one oxygen than to the other, and is covalently bound to the nearer oxygen. Because of its $\delta+$ charge, the hydrogen is attracted by lone pair electrons of the second oxygen. The hydrogen exchanges (moves back and forth) fast between the two positions O—H $\cdots$ O and O $\cdots$ H—O, but in ice at low temperatures the asymmetry of the grouping is evident. Because of the geometry of the arrangement, the polarity of the O—H bond, and so on, the hydrogen bond is considered to be a simple electrostatic interaction rather than a form of covalent bonding.

In a hydrogen bridge as in B_2H_6, the hydrogen is not on the line of centres of the bridged atoms, and is equidistant from them (B $\overset{H}{\diagup\diagdown}$ B). It is covalently bonded to both of them, by a three-centre bond, in which one electron pair is shared by all three atoms. The hydrogen bridge is much stronger (with respect to thermal dissociation) than are hydrogen bonds.

SAQ 15   Because the bridge bonding is rather weak (due to four atoms being held together by only two electron pairs) diborane readily rearranges on heating (in absence of air) to form other boranes and hydrogen; and it reacts very exothermically with Lewis bases such as water, or ammonia, to form the relatively strong N—O and B—N bonds.

SAQ 16   The theory of valence shell (electron pair) repulsion, as described in Unit 7, Section 7.4.3, and if applied to elemental carbon, would predict the tetrahedral coordination that is found in diamond. But graphite is the thermodynamically more stable form at normal temperatures and pressures, and electron repulsion in the valence shell of graphite must be minimal.

In graphite, each carbon uses its four valence electrons to form three σ bonds in the plane, and one π bond. Thus its octet consists of three σ-bonding electron pairs in the plane, trigonally disposed ($\widehat{CCC} = 120°$), and one electron pair in the bonding π orbital, which has the plane of carbons as its nodal plane. We can visualize this as (on average) one π electron above the plane and one below. See structure right.

This electron distribution has the symmetry of a trigonal bipyramid, the shape generated by five repulsion axes. The bond pairs at 120° in the plane repel each other less than do bond pairs at the tetrahedral angle 109.5°; and although repulsion is relatively high at the 90° angle between the axial and trigonal positions, this repulsion is in fact reduced by there being only (the equivalent of) one π electron at each axial position.

SAQ 17 Diamond does not conduct electricity because all the valence electrons are firmly localized in σ bonds. Diamond and graphite do not melt because their structures are held together by strong covalent bonds (in two dimensions in graphite) which can only be broken at very high temperatures. When they are broken, the fragments (C_2 etc.) are small and form a vapour, and when this condenses polymerization occurs to form the solid.

Graphite is a conductor of electricity because electrons are mobile within the giant π orbitals. It is used for electrodes because it is unreactive, insoluble, cheap, and can be made in robust form.

SAQ 18 Carbon in CO_2 shares two electrons with each oxygen to form two σ and two π bonds. There are no lone pairs on carbon, and the molecule must be linear. The bond order is two in each bond. (The bond length is, however, less than would be expected for

a structure $O=C=O$, and some resonance with forms such as $\overset{+}{O}\equiv C-\overset{-}{O}$ has been suggested, as in Evans, p. 159.)

CO_2 is isoelectronic with allene, $H_2C=C=CH_2$, and so the π orbitals for CO_2 are as given in Figure 33 for allene. These are sometimes called 'double streamers', the streamer on one π bond in the molecule being at 90° to the streamer on the other.

SAQ 19 Lewis acids include: BH_3, BF_3, BCl_3, BBr_3, $B(CH_3)_3$, $B(OH)_3$, $Be(OH)_2$, H^+ etc. Lewis bases include:

$$NH_3, CH_3NH_2, (CH_3)_2NH, (CH_3)_3N,$$
$$H_2O, CH_3OH, (CH_3)_2O,$$
$$F^-, Cl^-, Br^-, I^-, OH^-, O^{2-}$$

We study these further in Unit 11, Appendix 1.

SAQ 20

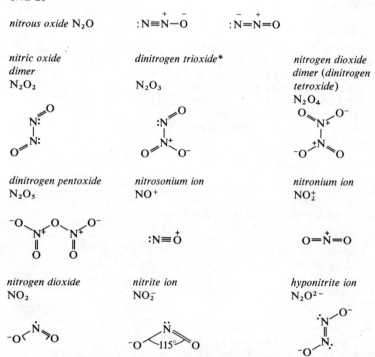

The nitrite ion is bent (as is the hyponitrite ion) because there is a lone pair on nitrogen. The angle in the NO_2 molecule is larger as there is only one electron in the lone pair orbital on nitrogen. In NO_2^+ there is no lone pair on nitrogen, and the ion is linear, as are allene and CO_2, with which NO_2^+ is isoelectronic.

This SAQ is an exercise in the valence bond method. The molecular orbital theory elegantly describes the bonding in these molecules and ions. They are all either linear or flat, with π orbitals covering the whole molecule: compare the description of nitrate ion in Section 9.3.

* This is the more stable form. The one given in Unit 6, SAQ 15, is a metastable form, found only at very low temperatures.

SAQ 21 This question can be answered in several ways. One could say that a hydrogen-ation catalyst would catalyse the decomposition of H_2O_2 (as in this Unit's TV programme) if any were formed. In fact, direct reaction of hydrogen and oxygen forms water, and not hydrogen peroxide, as the heat of reaction to form strong O—H bonds breaks the weak O—O bond. H_2O_2 is made indirectly, by hydrolysis of some other peroxide, under mild conditions.

SAQ 22 A second-row element, compared with later elements of its Group:

   (i)   is much smaller (as atom or ion; the CN is often smaller);
  (ii)   is more electronegative;
 (iii)   has a higher ionization energy;
  (iv)   readily forms $p\pi$ bonds (from boron in the row);
   (v)   is unable to expand its octet;
  (vi)   in Group II, is amphoteric, while the rest are not;
 (vii)   in Group III is a non-metal, the rest being metals;
(viii)   in Groups IV–VI, the stable form has a π-bonded structure, the rest not;
  (ix)   in Group V and VI, is a diatomic gas, the rest being solids, etc.

SAQ 23

1 Be, b	5 BBr_3, c	9 NO_2, c	13 K_2O, a
2 $(BeCl_2)_n$, d	6 SiC, d	10 N_2H_4, c	14 K_2O_2, e
3 B, d	7 CaC_2, e	11 $K_2B_2O_5$, e	
4 LiBr, a	8 H_3BO_3, c or d	12 $Mg(NO_3)_2$, e	

SAQ 24 A hydrogen-bond is a relatively weak interaction, as described in Unit 8, Section 8.8.1. Hydrogen-bonding in water, for example, is readily disrupted by a Lewis acid A to form AOH^-, or by a Lewis base B to form BH^+. Examples are:

$$B(OH)_3 + H_2O = B(OH)_4^- + H^+(aq)$$
$$NH_3 + H_2O = NH_4^+ + OH^-$$

A hydrogen bridge, though much stronger than hydrogen-bonding, is relatively weak as bonds go. Being electron-deficient, it is disrupted by a Lewis base to give an adduct, e.g.

$$B_2H_6 + 2(CH_3)_3N = 2(CH_3)_3N\!-\!BH_3$$

but the reaction is usually a vigorous one, with rearranged products.

SAQ 25 $\Delta H_f(M_{aq}^+) = \Delta H_{atm}^\ominus(M) + I_1(M) + \Delta H_{hyd}^\ominus(M) - [\Delta H_{hyd}^\ominus(H^+) + I_1(H) + \frac{1}{2}D(\text{H-H})]$

where $\Delta H_{hyd}^\ominus(H^+) + I_1(H) + \frac{1}{2}D(\text{H + H}) = 442\ \text{kJ mol}^{-1}$,
obtained from the values

$\Delta H_{hyd}^\ominus(H^+) = 1\ 090\ \text{kJ mol}^{-1}$ as above,

$I_1(H) \qquad = 1\ 314\ \text{kJ mol}^{-1}$ (*Data Book*, Section 9.1)

$\frac{1}{2}D(\text{H-H}) \quad = 218\ \text{kJ mol}^{-1}$ (*Data Book*, Section 8)

$\Delta H_{hyd}^\ominus(M) \quad = \Delta H_f^\ominus(M_{aq}^+) - [\Delta H_{atm}^\ominus(M) + I_1(M) - 442]$

where all measurements are in kJ mol^{-1}

Table 9   Determination of $\Delta H_{hyd}^\ominus$ for the alkali metal cations

$\Delta H_f^\ominus(M_{aq}^+)$	$\Delta H_{atm}^\ominus + I_1$	$\Delta H_{atm}^\ominus + I_1 -442$	$\Delta H_{hyd}^\ominus(M)$
kJ mol^{-1}	kJ mol^{-1}	kJ mol^{-1}	kJ mol^{-1}
	Data Book, Section 7		
Li —278	161 + 519	238	—516
Na —240	108 + 494	160	—400
K —252	90 + 418	66	—318
Rb —246	82 + 401	42	—288
Cs —248	78 + 377	13	—261

Thus the enthalpies of hydration become smaller (less negative) down the Group, as the ions become larger, and the solvating dipoles are farther from the centre of charge.

Answer to Exercise 1

The orbital occupancy in C_2 is shown in Figure 34. For C_2 there are six bonding electrons and two antibonding electrons. The net excess of bonding electrons is four, correspond-ing to a double bond. The C_2 molecule has been observed in flames; it is responsible for the blue colour in gas flames (S100, Unit 6, TV programme). From the spectroscopic data, we find that the bond is stronger than any of the single bonds we have encountered so far, for example, in Li_2, B_2 and H_2. The bond length is correspondingly smaller than that of B_2. Again the predictions are in agreement with spectroscopic evidence, which also shows that the highest energy electrons in C_2 are in π bonding orbitals.

The occupied orbitals in N_2 are also shown in Figure 34. With an excess of six bonding electrons, N_2 has an order of three, a triple bond. This is reflected by the very high bond dissociation energy (the N≡N triple bond is one of the strongest chemical bonds) and the correspondingly short bond length. Again, spectroscopic evidence supports the energy level diagram in Figure 34; the highest energy electrons in the nitrogen molecule are in the $\sigma 2p$ orbital.

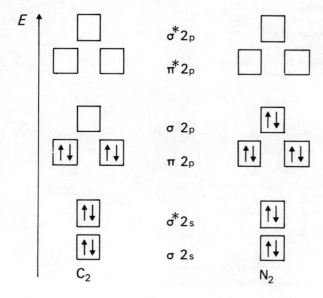

Figure 34 Orbital occupancies in C_2 and N_2.

NOTES

NOTES